METALS:

THEIR

PROPERTIES AND TREATMENT.

BY

CHARLES LOUDON BLOXAM,

Professor of Chemistry in King's College, London; and in the Royal Military Academy and the Department of Artillery Studies, Woolwich.

FOURTH EDITION.

NEW YORK:
D. APPLETON AND COMPANY,
549 & 551 BROADWAY.
1876.

Now in course of publication, in small 8vo. each volume containing about Three Hundred pages, bound in cloth,

A SERIES OF

ELEMENTARY WORKS ON MECHANICAL AND PHYSICAL SCIENCE,

FORMING A SERIES OF

TEXT-BOOKS OF SCIENCE,

ADAPTED FOR THE USE OF ARTISANS AND OF STUDENTS IN PUBLIC AND SCIENCE SCHOOLS.

The first Thirteen Works of the Series edited by T. M. GOODEVE, M.A. Lecturer on Applied Mechanics at the Royal School of Mines; and the remainder edited by C. W. MERRIFIELD, F.R.S. an Examiner in the Department of Public Education, and late Principal of the Royal School of Naval Architecture and Marine Engineering, South Kensington.

THE books of this Series are intended to serve for the use of practical men, as well as for exact instruction in the subjects of which they treat; and it is hoped that while retaining that logical clearness and simple sequence of thought which are essential to the making of a good scientific treatise, the style and subject-matter will be found to be within the comprehension of working men, and suited to their wants. The books will not be mere manuals for immediate application, nor University text-books, in which mental training is the formost object; but are meant to be *practical treatises, sound and exact in their logic, and with every theory and every process reduced to the stage of direct and useful application, and illustrated by well selected examples from familiar processes and facts.* It is hoped that the publication of these books—in addition to other useful results—will tend to the leading up of Artisans to become candidates for the WHITWORTH SCHOLARSHIPS.

TEXT-BOOKS, NOW PUBLISHED, EDITED BY T. M. GOODEVE, M.A.

THE ELEMENTS OF MECHANISM.

Designed for Students of Applied Mechanics. By T. M. GOODEVE, M.A. Lecturer on Applied Mechanics at the Royal School of Mines, and formerly Professor of Natural Philosophy in King's College, London; Joint-Editor of the Series. New Edition, revised; with 257 Figures on Wood. Price 3*s.* 6*d.*

'The object of the present series of convenient and elegant *Text-Books of Science* is somewhat peculiar, for they are intended to occupy an intermediate place between Art and Science. They are neither mere manuals for immediate application on the one hand, nor on the other University text-books, in which mental training is the foremost object. They explain principles and give scientific methods, but only just so far as it is necessary for practical application, and they illustrate this application by a great number of familiar examples. Similar works have been attempted before, but for the most part in a very rough and coarse way. The speciality of this series consists in the fact that men of the highest scientific eminence in their respective departments have been engaged to write them; so that the books, while not professing to exhaust their subjects, and being, in fact, definitely confined within certain limits, will nevertheless be perfectly sound and exact as far as they go, and may at any time be made the basis for going farther. Three of them, which lie before us, fully justify this description. *Algebra and Trigonometry*, by the Rev. W. N. GRIFFIN, is a concise and clearly arranged treatise. *The Elements of Mechanism*, by T. M. GOODEVE, is a very full description of all the ingenious methods by which one form of motion is converted into another. Cranks and rods and toothed wheels, escapements and fusees, are made as plain as pen and pencil can make them. *Inorganic Chemistry*, by the late Professor MILLER, whose recent death is a great loss to chemical science, is treated in a remarkably clear and simple style. Two objects have been kept in view in these *Text-Books*, one general and the other particular. They are meant to help artisans in self instruction, and to lead up to the Whitworth Scholarships. But they will be found very useful in Schools also.'

GUARDIAN.

METALS, THEIR PROPERTIES AND TREATMENT.

By CHARLES LOUDON BLOXAM, Professor of Chemistry in King's College, London; Professor of Chemistry in the Department of Artillery Studies, and in the Royal Military Academy, Woolwich. With 105 Figures on Wood. Price 3*s.* 6*d.*

'Hitherto text-books on metallurgical science have either been of so large and expensive a nature, like the splendid and exhaustive work of Dr. PERCY, or they have been mere flimsy pretences, of little or no value to the real student. In the first case, they are both too expensive and too difficult for the beginner; and in the latter, the harm they do by describing exploded processes and antiquated operations is much greater than any good they can do to the reader. Professor BLOXAM, with a thorough practical knowledge of his subject, has also another quality of the greatest value to writers of books like the one before us—he can describe with great simplicity and clearness the various operations and the apparatus employed, and the construction of the works in which they are carried on. This is the result of being personally conversant with them, and it gives to the book an interest to the general reader as well as to the student in technical metallurgy. The subject itself is one of immense interest and every intelligent person must feel pleasure in gaining some knowledge of the various, and in many instances wonderful, processes by which man wins from the earth the precious and useful metals, and converts them to his use in almost numberless ways. With such a manual as this no difficulty will be felt in gaining such knowledge, and we feel sure it will, ere long, be in the hands of many who have no intention of practically pursuing the metallurgic art. To the student it will supply all the knowledge necessary for primary examinations; and will, by the clear descriptions and excellent diagrams and woodcuts, convey to him very comprehensive information as to the construction of the most improved furnaces for smelting and refining works, together with the most recent improvements in apparatus and chemical processes employed both in this country and abroad.' SCOTSMAN.

INTRODUCTION TO THE STUDY OF INORGANIC CHEMISTRY.

By WILLIAM ALLEN MILLER, M.D. LL.D. F.R.S. late Professor of Chemistry in King's College, London; Author of 'Elements of Chemistry, Theoretical and Practical.' New Edition, revised; with 71 Figures on Wood. 3*s.* 6*d.*

'This text-book of inorganic chemistry is one of the most useful elementary manuals we have met with for a long time.' PHILOSOPHICAL MAGAZINE.

ALGEBRA AND TRIGONOMETRY.

By the Rev. WILLIAM NATHANIEL GRIFFIN, B.D. sometime Fellow of St. John's College, Cambridge. Price 3*s.* 6*d.*

'We have examined this volume with much care. From our previous knowledge of Mr. GRIFFIN'S antecedents and writings, we were led to expect accuracy of reasoning and mathematical precision. We have not been disappointed. While explaining in a scientific yet popular style the rudiments of the two subjects of which he has here treated, Mr. GRIFFIN has not lost sight of some of the difficulties that meet the learner at the beginning and during the progress of his investigations. These, however, he has so explained as to make them rather an attraction than a hindrance. We have no doubt that students, whether artisans or others, who fairly master this book, will have imbibed such an acquaintance with the subjects it explains as will induce them to prosecute their investigations into those questions which are being continually raised in connection with the applied sciences.' NATIONAL SOCIETY'S PAPER.

NOTES ON THE ELEMENTS OF ALGEBRA AND TRIGONOMETRY;

With SOLUTIONS of the more difficult QUESTIONS. By the Rev. WILLIAM NATHANIEL GRIFFIN, B.D. sometime Fellow of St. John's College, Cambridge. Price 3*s.* 6*d.*

PLANE AND SOLID GEOMETRY.

By the Rev. H. W. WATSON, formerly Fellow of Trinity College, Cambridge, and late Assistant-Master of Harrow School. Price 3*s.* 6*d.*

'The book may be safely used as an excellent manual of geometry; and we think that, on the whole, we should prefer to put it rather than any edition we know of *Euclid* into the hands of a person attempting unaided to acquire the elements of the science.' ATHENÆUM.

'It is altogether a very practical and well-arranged treatise on geometry; and we recommend it to the attention of teachers who are desirous of replacing the time-honoured *Elements of Euclid* by a text-book more in harmony with the present state of mathematical science.' EDUCATIONAL TIMES.

THEORY OF HEAT.

By J. CLERK MAXWELL, M.A. LL.D. Edin. F.R.S.S. L. & E. Professor of Experimental Physics in the University of Cambridge. New Edition, revised; with 41 Woodcuts and Diagrams. Price 3*s.* 6*d.*

'Considered as addressed to students already well trained in something more than the elements of mathematics, and familiar with the fundamental laws of mechanics, it would be hard to name a better book. To such readers it will prove an excellent introduction to the very difficult science of Thermodynamics. They will find in it, written by a master, an admirable account of the existing state of knowledge as to the nature and effects of heat, of the steps by which that knowledge has been acquired, of its bearing on the molecular constitution of matter, and of the numerous points at which the subject of heat touches the general doctrines of mechanics.'

PHILOSOPHICAL MAGAZINE.

TECHNICAL ARITHMETIC AND MENSURATION.

By CHARLES W. MERRIFIELD, F.R.S. an Examiner in the Department of Public Education; Joint-Editor of the Series. Price 3*s.* 6*d.*

'Notwithstanding that arithmetic has formed the subject of innumerable treatises by all classes of writers, from the schoolmaster, anxious to advertise his suburban academy for young gentlemen, to the mathematician whose reputation is of world-wide renown, until every method of treating the theme would appear to have been pretty well exhausted, Mr. Merrifield has succeeded in producing a thoroughly original work. The task which he had set himself in writing the book, was, he tells us in his preface, to give "to the elementary rules the precision and illustration which they need for the further pursuit of the subject, and to the higher rules that gradual induction which is a more effective instrument of teaching than a strict logical arrangement." It must be admitted that he has fully succeeded in his object. It is almost needless to say that this book is got up in the careful manner which distinguishes others of the same series; while its low price will render it especially welcome both to the teachers of science classes and to the mechanic who is devoting his leisure hours to self-tuition. Considering the enormous development which technical instruction is just now receiving, we venture to anticipate for it an extremely large sale.'

ENGINEER.

KEY TO MERRIFIELD'S TEXT-BOOK OF TECHNICAL ARITHMETIC AND MENSURATION.

By the Rev. JOHN HUNTER, M.A. one of the National Society's Examiners of Middle-Class Schools; formerly Vice-Principal of the National Society's Training College, Battersea. Price 3*s.* 6*d.*

THE STRENGTH OF MATERIALS AND STRUCTURES.

The Strength of Materials as depending on their quality and as ascertained by Testing Apparatus; the Strength of Structures, as depending on their form and arrangement, and on the Materials of which they are composed. By JOHN ANDERSON, C.E. LL.D. F.R.S.E. Superintendent of Machinery to the War Department. Price 3*s.* 6*d.*

ELECTRICITY AND MAGNETISM.

By FLEEMING JENKIN, F.R.S.S. L. & E. Professor of Engineeringin the University of Edinburgh. New Edition, revised. Price 3*s.* 6*d.*

WORKSHOP APPLIANCES.

Including Descriptions of the Gauging and Measuring Instruments, the Hand Cutting-Tools, Lathes, Drilling, Planing, and other Machine Tools used by Engineers. By C. P. B. SHELLEY, Civil Engineer, Hon. Fellow and Professor of Manufacturing Art and Machinery at King's College, London. With 209 Figures on Wood. Price 3*s.* 6*d.*

PRINCIPLES OF MECHANICS.

By T. M. GOODEVE, M.A. Barrister-at-Law, Lecturer on Applied Mechanics at the Royal School of Mines; Joint-Editor of the Series. With 208 Figures and Mathematical Diagrams engraved on Wood. Price 3*s.* 6*d.*

TEXT-BOOKS, NOW PUBLISHED, EDITED BY C. W. MERRIFIELD, F.R.S.

INTRODUCTION TO THE STUDY OF ORGANIC CHEMISTRY;

The CHEMISTRY of CARBON and its COMPOUNDS.

By HENRY E. ARMSTRONG, Ph.D. F.C.S. Professor of Chemistry in the London Institution. With 8 Figures on Wood. Price 3*s.* 6*d.*

QUANTITATIVE CHEMICAL ANALYSIS.

By T. E. THORPE, Ph.D. F.R.S.E. Professor of Chemistry in the Yorkshire College of Science, Leeds. With 88 Figures on Wood. Price 4*s.* 6*d.*

QUALITATIVE CHEMICAL ANALYSIS AND LABORATORY PRACTICE.

By T. E. THORPE, Ph.D. F.R.S.E. Professor of Chemistry in the Yorkshire College of Science, Leeds; and M. M. PATTISON MUIR, F.R.S.E., Professor of Chemistry in the Andersonian University, Glasgow. With Plate and 57 Figures on Wood. Price 3*s.* 6*d.*

TEXT-BOOKS PREPARING FOR PUBLICATION, TO BE EDITED BY
T. M. GOODEVE, M.A.

ECONOMICAL APPLICATIONS OF HEAT.

Including Combustion, Evaporation, Furnaces, Flues, and Boilers.

By C. P. B. SHELLEY, Civil Engineer, and Professor of Manufacturing Art and Machinery at King's College, London.

With a Chapter on the Probable Future Development of the Science of Heat, by C. WILLIAM SIEMENS, F.R.S.

THE STEAM ENGINE.

By T. M. GOODEVE, M.A. Lecturer on Mechanics at the Royal School of Mines and formerly Professor of Natural Philosophy in King's College, London Joint-Editor of the Series.

TEXT-BOOKS PREPARING FOR PUBLICATION, TO BE EDITED BY
C. W. MERRIFIELD, F.R.S.

RAILWAY APPLIANCES,

Including Permanent Way, Points and Crossings, Stations and Station Arrangements, Signals, Carriage and Wagon Stock, Breaks, and other Details of Railways. By JOHN WOLFE BARRY, Member of the Institution of Civil Engineers, &c.

TELEGRAPHY.

By W. H. PREECE, C.E. Divisional Engineer, Post Office Telegraphs; and J. SIVEWRIGHT, M.A. Superintendent (Engineering Department) Post Office Telegraphs.

PRACTICAL AND DESCRIPTIVE GEOMETRY, AND PRINCIPLES OF MECHANICAL DRAWING.

By C. W. MERRIFIELD, F.R.S. an Examiner in the Department of Public Education; Joint-Editor of the Series.

ELEMENTS OF MACHINE DESIGN.

With Rules and Tables for Designing and Drawing the Details of Machinery. Adapted to the use of Mechanical Draughtsmen and Teachers of Machine Drawing.

By W. CAWTHORNE UNWIN, B. Sc. Assoc. Inst. C.E. Professor of Hydraulic and Mechanical Engineering at Cooper's Hill College.

PHYSICAL GEOGRAPHY.

By the Rev. GEORGE BUTLER, M.A. Principal of Liverpool College; Editor of 'The Public Schools Atlas of Modern Geography.'

*** *To be followed by other Works on other Branches of Science.*

London, LONGMANS & CO.

TEXT-BOOKS OF SCIENCE

ADAPTED FOR THE USE OF

ARTISANS AND STUDENTS IN PUBLIC AND OTHER SCHOOLS.

METALS.

LONDON: PRINTED BY
SPOTTISWOODE AND CO., NEW-STREET SQUARE
AND PARLIAMENT STREET

PREFACE.

THE following treatise has been written for the Series of Text-Books of Science published by Messrs. Longmans and Co., and intended for the use of Artisans and Students in Public and other Schools.

Its subject, Metallurgy, one of the most ancient of the Arts, and closely connected with the progress of civilization, has attained to enormous proportions. The processes followed in the treatment of any particular metal, or of its ores, are varied according to local circumstances and traditions; a few only, of modern origin, having been devised upon scientific principles. To describe these processes, with all their modifications, would require several large volumes. The Author, consulting the interests of those persons for whom the Text-Books are designed, has endeavoured to give such a description of the mode of dealing with the useful metals as shall enable the Chemical principles involved to be clearly understood, and shall prepare the reader for the more detailed descriptions which are to be found in larger works.

The Student is recommended to test his knowledge by the Questions at the end of the book.

WOOLWICH:
November 1872.

CONTENTS.

METALS:

THEIR PROPERTIES AND TREATMENT.

THE METALS.

THE word METAL appears to be derived from the Greek *μετ' ἄλλα*, *in quest of other things*, whence come *μεταλλάω*, *to search after*, *to explore*, or, in gold-diggers' language, *to prospect*, and the corresponding substantive *μέταλλον*, *a mine*.

About fifty of the undecomposed or *elementary* substances are classed together under the head of METALS by the chemist, because they manifest certain properties when acted upon by chemical tests, without regard to those external characters which are commonly associated with the idea of a metal.

Many of these are unfit to be employed in the metallic state for any of the ordinary uses of metals, because they cannot be exposed to the action of air, even for a short time, without being rusted or corroded, by combining with the oxygen of the air, to such an extent that they entirely lose their metallic characters.

Among those which offer sufficient resistance to the action of air, many are excluded from useful application in their metallic state, on account of their rarity, or of the great difficulty which is experienced in extracting them from their ores.

The metals which are employed for useful purposes in their pure or metallic state are—

Aluminum	Copper	Magnesium	Platinum
Antimony	Gold	Mercury	Silver
Bismuth	Iron	Nickel	Tin
Cadmium	Lead	Palladium	Zinc

On considering this list, it will be seen that several of the metals named in it are employed to produce some effect dependent upon a peculiar property of the metal, and not upon qualities which belong to it in common with the rest. Thus, mercury or quicksilver is used for *amalgamating* or dissolving other metals, and also as a suitable liquid for constructing barometers and thermometers; antimony owes its usefulness to its property of hardening lead and tin when melted with them; bismuth and cadmium are employed to render lead and tin capable of being melted at lower temperatures; nickel is used to whiten copper in order to make German silver; and magnesium is valuable for its property of burning easily with production of a brilliant white light.

Moreover, gold, platinum, palladium, and silver, being comparatively rare, and aluminum being obtainable by a somewhat costly process, the useful applications of these metals are limited by their high price, so that there remain only TIN, LEAD, COPPER, IRON and ZINC to be considered as metals largely employed for useful purposes.

The qualities possessed by these metals, rendering them fit for purposes which could not be fulfilled by non-metallic substances, are *lustre*, or the power of reflecting light; *tenacity*, or resistance to any attempt to pull asunder their particles; *malleability*, or the capability of being hammered or rolled into thin sheets; *ductility*, or the property of being drawn out into wire; high *specific gravity*, or relative weight; high *conducting power* for heat and electricity; and *fusibility*, or the property of becoming liquid when heated.

METALLIC LUSTRE.—The power of reflecting the rays of light is possessed in a much higher degree by metals than by non-metallic substances. Although some examples of the latter class, such as iodine and plumbago, reflect much of the light which falls upon smooth surfaces of them, they have a black appearance, caused by their absorbing a large proportion of the luminous rays, which is quite different from the true metallic lustre. Iron, in the form of steel, is capable of exhibiting this lustre in very great perfection, because the hardness of steel allows its surface to be ground perfectly smooth by the application of fine particles of very hard substances, such as emery and diamond-dust, which rub off minute projections from the surface without producing scratches or indentations. A surface so polished sends back directly to the eye of the observer almost all the light falling upon it, whilst a rough surface, being made up of a number of small surfaces, scatters the reflected rays in all directions. Tin is naturally a brilliant metal, but is not hard enough to be polished, like steel; if, however, it be dissolved in twice its weight of melted copper, an *alloy* of great hardness and brilliancy is formed, which is employed for the *specula* or mirrors of reflecting telescopes. Zinc and lead exhibit the metallic lustre in an inferior degree, and become dull when exposed to air, because the metal at the surface combines with oxygen, forming a thin film of oxide which has no metallic lustre. The natural lustre of silver is very great, and, if it be hardened by admixture with a little copper, it becomes susceptible of a very high polish which is not dimmed by the action of the oxygen of air, though it is easily tarnished by sulphur existing in foul air in the form of sulphuretted hydrogen. The splendid combination of lustre and colour exhibited by burnished gold is proverbial, and is undiminished by the action of the atmosphere. The lustre of palladium and platinum resembles that of silver, and is not affected by oxygen or sulphur in the air. Aluminum has also a permanent lustre, though inferior to that of silver. When dissolved in nine times its weight of melted copper, aluminum forms a hard yellow alloy capable of being polished to resemble gold, but becoming slowly tarnished by the action of the oxygen in air.

TENACITY.—The strength with which the metals oppose any attempt to pull asunder their particles is one of their most useful properties, and is determined by ascertaining the exact weight which must be suspended from the ends of wires or rods of equal diameter, in order to break them. The weight required to break a given metallic wire is found to vary according to the manner in which the strain is applied, the resistance of the wire being greater when the whole of the breaking weight is applied at once than when it is added gradually, probably because, in the latter case, the wire becomes stretched and weakened by each additional weight.

Steel (iron combined with about $\frac{1}{100}$th part of carbon) is by far the most tenacious of metals, and lead is the least tenacious of those in ordinary use.

If the weight required to pull asunder a wire of lead be taken as unity, that required by similar wires of the other metals will found to approach nearly to the numbers contained in the following table :—

Relative Tenacity of the Metals.

Metal	Tenacity	Metal	Tenacity
Lead	1	Silver	12½
Tin	1⅓	Platinum	15
Zinc	2	Copper	18
Palladium	11½	Iron	27½
Gold	12	Steel	42

The tenacity of metals is very seriously affected by variations in their structure, purity and temperature. Thus, rods of metal which have been cast in a mould are generally weaker than rods of equal dimensions made by drawing the metal through the gradually diminishing holes of the wire-drawer's plate. The tenacity of iron rods which have been rolled until they have acquired a fibrous structure, is much higher than that of rods which are *crystalline* in texture, the metal tending to break asunder where the smooth surfaces of the separate crystals are in contact with each other.

The tenacity of a metal when hot is, as might be anticipated, less than its tenacity when cold; and if the metal be made red-hot and allowed to cool slowly, it will generally be found to have diminished in tenacity, probably because a high temperature tends to encourage the formation of a crystalline structure. The effect of the presence of impurities upon the tenacity of metals will be more appropriately studied when the individual metals are under consideration, but it may be stated generally that chemical purity is not of necessity accompanied by the highest degree of tenacity. Thus, the small proportion of carbon present in steel is seen in the above table to have greatly increased the tenacity of the iron, and pure zinc has a much lower tenacity than the ordinary zinc of commerce.

MALLEABILITY.—The facility with which a mass of metal can be hammered or rolled into a thin sheet without being torn, must depend partly upon its softness, and partly upon its tenacity. If it depended upon softness alone, lead should be the most malleable of ordinary metals; but, although it is easy to hammer a mass of lead into a flat plate, or to squeeze it between rollers, any attempt to reduce it to an extremely thin sheet fails from its want of tenacity, which causes it to be worn into holes by percussion or friction. On the other hand, if malleability were entirely regulated by tenacity, iron would occupy the first place, whereas, on account of its hardness, it is the least malleable of metals in ordinary use; whilst gold, occupying an intermediate position with respect to tenacity, is the most malleable, which appears surprising to those who are only acquainted with gold in its ordinary forms of coin and ornament, in which it is hardened and rendered much less malleable by the presence of copper and silver.

During the rolling or *lamination* of metals their particles are obviously squeezed into unnatural positions; it becomes necessary, therefore, in order to avoid breaking, to enable the particles to resume their former relative situations; this is effected by heating the metallic sheet after every two or three rollings, and allowing it to cool slowly, a process of *annealing* similar to that by which glass vessels are rendered less brittle.

In the following table the ordinary metals are arranged in the order of malleability :—

Table of Malleability.

1. Gold.	4. Tin.	7. Zinc.
2. Silver.	5. Platinum.	8. Iron.
3. Copper.	6. Lead.	

Ductility.—The ease with which a metal can be elongated into a wire, by being drawn through the gradually diminishing holes of the wire-drawer's plate, will be greater in proportion to the softness of the metal; but the thinness of the wire to which it can be reduced is regulated by the tenacity of the metal, which enables it to resist, without breaking, the force required to draw it through the holes. And it is found that their tenacity has more influence upon the ductility of metals than upon their malleability, for the particles of a weak metal, like tin, may cohere under the hammer, although they would be easily torn apart by the direct pull necessary in wire-drawing.

Gold, silver, and platinum, which occupy an intermediate position with respect to tenacity, are the most ductile of the metals, whilst tin and lead, which are lowest in tenacity, are the least ductile, though their softness gives them a higher place in the order of malleability.

Table of Ductility.

1. Gold.	5. Copper.	8. Zinc.
2. Silver.	6. Palladium.	9. Tin.
3. Platinum.	7. Aluminum.	10. Lead.
4. Iron.		

The metals require annealing during the process of wire-drawing, as in that of lamination, and for a similar reason.

Specific Gravity.—The relative weights of equal bulks of the metals exercise considerable influence upon their useful applications. The relative weight of gold being very high it is well adapted for a circulating medium, a large value being compressed into a portable form. On the other hand, iron would be employed with far less advantage in building if its relative weight did not happen to be low, whilst aluminum, being the lightest of metals in ordinary use, is particularly well adapted for the production of small weights, as fractions of a grain, which shall yet be large enough to handle; such weights being nearly nine times as large when made of aluminum as they are when platinum is employed, as was the case before the introduction of aluminum.

The *specific gravities*, or comparative weights of equal bulks of the metals, are generally expressed by numbers which show that each metal is so many times as heavy as an equal bulk of pure distilled water at the ordinary temperature (60° F.); thus, zinc is a little more than seven times as heavy as an equal bulk of water, so that its specific gravity is expressed by 7 and a fraction.

The first column of numbers in the following table gives the specific gravities of the metals in round numbers, which can be easily retained in the memory, and are sufficiently exact for ordinary purposes, the more accurate numbers usually employed in scientific works being given in the next column :—

Table of Specific Gravities of the Metals.

Platinum	$21\frac{1}{2}$	21·53	Nickel	$8\frac{4}{5}$	8·82
Gold	$19\frac{1}{3}$	19·34	Iron	$7\frac{4}{5}$	7·84
Mercury	$13\frac{3}{5}$	13·59	Tin	$7\frac{1}{3}$	7·29
Palladium	$11\frac{4}{5}$	11·8	Zinc	$7\frac{1}{7}$	7·14
Lead	$11\frac{1}{3}$	11·36	Antimony	$6\frac{2}{3}$	6·71
ilver	$10\frac{1}{2}$	10·53	Aluminum	$2\frac{2}{3}$	2·67
Bismuth	$9\frac{4}{5}$	9·79	Magnesium	$1\frac{3}{4}$	1·74
Copper	9	8·95			

CONDUCTING POWER OF METALS FOR HEAT. — The sensation of cold when the hand is placed upon a piece of metal of the ordinary temperature of the air shows us that metals are better conductors of heat than non-metallic bodies, for the particles of metal which are first warmed by contact with the hand give up the acquired heat to the neighbouring particles, and being thus cooled to nearly their former temperature, are able to abstract a fresh supply of heat from the hand; whereas, when the hand is placed upon wood, or other inferior conductors of heat, the particles in contact with it are warmed by the removal of a trifling amount of heat from the hand, and are not soon cooled again by parting with their heat to the particles adjoining. In consequence of the rapidity with which heat applied to one portion of a mass of metal is communicated to the whole of the particles composing it, metals may be suddenly heated or cooled with much less risk of causing them to crack or *fly* than is the case with non-metallic substances. When an earthenware pipkin or a glass bottle is placed upon the fire, the outside immediately becomes much hotter than the inside, and being expanded by the heat, tears apart the particles of the inside of the vessel and produces a crack, but in the case of a metallic vessel the heat is rapidly transmitted, and all parts of the vessel are expanded almost simultaneously. The much greater rapidity with which water can be heated in metallic vessels is another useful result of the superior conducting power of the metals.

In the following table the metals are arranged in the order of their conducting power, the first being the best conductor :—

Table of Conducting Power for Heat.

1. Silver.	5. Zinc.	9. Lead.
2. Gold.	6. Iron.	10. Antimony.
3. Copper.	7. Tin.	11. Bismuth.
4. Aluminum.	8. Platinum.	

CONDUCTING POWER OF METALS FOR ELECTRICITY.—The conducting power for electricity, of metals, refers to the facility with which an electric disturbance excited in one portion of a mass of metal is transmitted to the other particles composing the mass. Thus, a very slight electric disturbance at one end of a copper wire is sufficient to produce movement in a telegraph needle at the other extremity, whilst a much greater amount of disturbance, or, in other words, a more powerful *current*, is required if an iron wire of the same length and thickness be employed.

Only one non-metallic substance—carbon, in some of its varieties—at all approaches to the metals in the power of conducting electricity.

Those metals which are the best conductors of heat are also the best conductors of electricity, and in both cases the conducting power is seriously impaired by the presence in the metal even of small quantities of other metals, or of non-metallic bodies, as well as by an increase of temperature in the metal. When heated to the boiling point of water, the metals have only about three-fourths of the conducting power which they exhibit at the freezing point.

The following table shows the relative conducting power of the most important metals, in a pure state, at 32° F., the conducting power of silver, which is higher than that of any other metal, being taken as 1000:—

Table of Conducting Power for Electricity.

Silver = 1000

Metal	Power	Metal	Power
Copper	999	Nickel	131
Gold	779	Tin	123
Zinc	290	Lead	83
Palladium	184	Antimony	46
Platinum	180	Bismuth	12
Iron	168		

Fusibility.—Although the property of becoming liquid at high temperatures is not confined to the metals, it must be mentioned among the properties which conduce to their utility, for it enables the founder to produce a large number of objects of a given pattern with little expenditure of time and labour, and offers to the worker in metals a ready method of soldering together, in a durable manner, the separate pieces of his work. Tin and lead, being the most fusible of ordinary metals, are the constituents of solder, whilst iron (wrought iron), as the least fusible of the common metals, is used for firebars, melting-pots, and similar purposes.

Table of Fusibility.

Tin	melts at	442° F.	Silver	melts at	1800° F.
Cadmium	,,	442	Copper	,,	1990
Bismuth	,,	507	Gold	,,	2000
Lead	,,	617	Cast Iron	,,	2780
Zinc	,,	773	Steel	,,	4000
Antimony*	,,	1150	Wrought Iron	,,	above 4000

Platinum melts only in the oxy-hydrogen blowpipe flame.

In practical work the temperature is commonly inferred from the appearance of the fire; thus, the *red heat* of an ordinary domestic fire is roughly valued at 1000° F., so that tin, lead, and zinc can be very easily melted in a crucible or ladle placed in such a fire; but aluminum, silver, copper, and gold, require a *bright red* (*cherry red*) or furnace heat to melt them; cast iron requires a very bright red heat, only attainable in a furnace with a very good draught; and for melting steel, a furnace of special construction (*wind furnace*) is employed. Wrought iron can be fused only at a white heat, producible by a blast of air in a forge, and platinum melts at a greenish white heat in the flame of hydrogen supplied with pure oxygen. The production of a temperature adequate to the fusion of steel and wrought iron in large quantities has been much facilitated by the introduction of Siemens' regenerative furnace, in which the waste heat of the fire, instead of escaping up the chimney into the air, is accumulated in masses of fire-brick, and restored again to the furnace.

* Estimates of temperature above the fusing point of zinc cannot be regarded as exact, on account of the difficulty of ascertaining them.

IRON.

With respect to its useful properties, iron occupies the first place among the metals. By far the strongest, and, at the same time, one of the lightest, its applications in the arts of construction are much more numerous than those of any other metal. Being capable of assuming, according to the treatment which it undergoes, the forms of wrought iron, cast iron, and steel, it is susceptible of the widest variations in its characters. Extracted from its ores in the form of cast iron, it is melted with comparative facility, and, according to the mode of operating in the foundry, may be made to yield castings which are easily filed and turned, or may be rendered so hard that no tool is able to touch it. By judicious treatment with heat and atmospheric air the cast iron is converted into steel, the strongest, and one of the hardest and most elastic of all materials, as well as the only one of which a magnetic needle can be made. Continued a little further, the joint action of heat and atmospheric air converts the steel into wrought iron, possessing great strength and toughness, yet soft enough to be turned, bored, and punched with ease, and, especially when heated, to be rolled and twisted into the most varied forms without cracking. With less disposition to melt under the action of heat than any other common metal, wrought iron is sufficiently softened at a bright red heat to be *welded* or joined to another piece in the most perfect manner, without the use of solder of any kind. Being capable of acquiring and of losing the properties of a magnet with great rapidity, soft iron (wrought iron) is the only material which is adapted for the construction of electro-magnetic and magneto-electric apparatus.

It is not too much to assert that scarcely a step of importance has ever been made in the industrial progress of any community to which some one of the three modifications of iron has not been indispensable.

Possessed of so many valuable qualities, iron is still the cheapest of all the metals, since the ores from which it is extracted are scattered in profusion through the crust of the earth, and can be made to yield the metal in abundance by a moderate expenditure of time, labour, and fuel.

ORES OF IRON.—Iron in the metallic condition, or *native iron*, is very rarely found in nature. Nearly all the specimens which have been examined have been *meteoric iron*, occurring in masses of irregular form, which have descended upon the surface of the earth, but whence they are derived is at present only a matter for speculation. Such masses have been found containing 93 parts in the hundred, of metallic iron, always associated with nickel, and sometimes with small quantities of other metals, and of phosphorus, sulphur and carbon. They vary much in size; two masses of iron, supposed to be of meteoric origin, have been recently found on the coast of Greenland, weighing, respectively, 21 tons and 9 tons. A small one, which was found at Lenarto in Hungary and weighed about 190 lbs., was remarkably malleable, and its analysis furnished the following results:—

Lenarto Meteoric Iron.

Specific Gravity, 7·79

Iron	90·883
Nickel	8·450
Cobalt	0·665
Copper	0·002
	100·000

A recent examination of this meteoric iron has led to the very interesting discovery that it contains about twice and a half its volume of hydrogen gas, apparently in an uncombined state.

Iron is most commonly found in a state of chemical combination with oxygen or sulphur, which disguise its metallic properties and convert it into earthy or stony masses. The compound of iron with oxygen, or *oxide of iron*, which is familiar to us in the form of *rust*, occurs in a very large number of mineral substances, and is often the cause of their colour. Sand, clay, and gravel, commonly owe their yellow, brown, or red shade to the presence of oxide of iron, a small proportion of which imparts a very distinct colour. No mineral substance, however, would be considered as an *ore* of iron which contained less than about twenty parts of iron in the hundred, for otherwise it would not repay the cost of its extraction.

The following table includes the mineral substances which are commonly regarded as ores of iron :—

Ores of Iron.

	Composition	Iron in 100 parts of pure Ore
Magnetic Iron Ore	Iron, Oxygen	72
Red Hæmatite	Iron, Oxygen	70
Specular Ore	Iron, Oxygen	70
Brown Hæmatite	Iron, Oxygen, Water	60
Spathic Iron Ore	Iron, Oxygen, Carbonic Acid	48
Clay Iron Stone	Iron, Oxygen, Carbonic Acid, Clay	Variable, 17 to 50
Black-band Ore	Iron, Oxygen, Carbonic Acid, Clay, Bituminous matter	Variable, 21 to 43
Iron Pyrites*	Iron, Sulphur	46

Magnetic Iron Ore, or *Magnetite*, derives its name from Magnesia in Asia Minor, where its power of attracting iron and steel was first observed. One variety of the ore constitutes the *loadstone*,† which confers magnetic properties upon

* Iron Pyrites is not worked as an ore of iron, on account of the great difficulty of separating all the sulphur from the metal.

† Probably corrupted from *lead-stone*, the stone which leads or guides, in allusion to its use for making the needle of the mariner's compass

steel. This variety occurs chiefly in Siberia and the Hartz. Some of the common varieties of the ore do not attract iron, although they are capable, like steel, of becoming magnetic when brought into contact with powerful magnets.

This ore is generally met with in compact heavy masses of an iron black or grey colour, and with considerable lustre. Its specific gravity varies from 4·9 to 5·2. It abounds chiefly in the northern parts of the globe, and is found in immense masses in Norway, Sweden, Russia, and North America, being the most important iron ore in those countries. The iron extracted from it is generally of excellent quality, though it is occasionally deteriorated by sulphur and phosphorus, which are derived respectively from iron pyrites and from *apatite* (phosphate of lime), which are sometimes found associated with the magnetic ore. The bulk of the Swedish iron, so much valued for the manufacture of steel, is extracted from the magnetic ore at Dannemora, where it is worked in an open quarry.

Magnetic iron ore often contains titanic acid, the oxide of the metal titanium, and this is especially the case with a variety of the ore which occurs as a heavy black shining sand in India, Nova Scotia, and New Zealand.

Red Hæmatite has been so called from the Greek word signifying *blood*, on account of its dark red colour,* and is sometimes erroneously called *bloodstone* (the true bloodstone being a dark green variety of silica (*heliotrope*) with red spots). In appearance it is the most striking of the ores of iron, sometimes appearing in rounded masses having externally a liver colour with considerable lustre, and internally made up of layers having the appearance of the thick shell of some huge fruit, or of bundles of fibres, which look like petrified wood. The specific gravity of this variety is

magnetic. In a similar way we have *load-star*, the star which leads towards the pole; *loadsman*, one who guides—a pilot; *load* or *lode*, the leading vein in a mine.

* The termination *ite*, so generally found in the names of minerals, was originally derived from the Greek word for *stone*.

about 5·0. Such specimens are in general remarkably hard, and are useful for burnishing metals. But it is occasionally found in much softer earthy-looking masses of a brighter colour, and not uncommonly associated with clay.

Red hæmatite is much more generally diffused than magnetite, and is found abundantly in this country, at Whitehaven (Cumberland) and Ulverstone (Lancashire), as well as in Glamorganshire. The compact variety is exceedingly pure, and furnishes iron of the very best quality, but its very dense structure renders it difficult to smelt alone in English furnaces, so that it is customary to mix it with some lighter ores of inferior purity, a practice which detracts from the excellence of the iron obtained.

This ore is also abundant in Ireland (Balcarry Bay), North America, Saxony, Bohemia, and the Hartz.

Specular Iron Ore is identical in composition with red hæmatite, though its appearance is so different, for it forms crystalline masses, and sometimes separate crystals of a steel-grey colour and brilliant lustre, whence it derives its name (*speculum* [Latin], a mirror). It is also called *Iron-glance* (*Glanz* [German], lustre), and in one of its varieties, made up of shining scales, *Micaceous Iron Ore* (*micare* [Latin], to shine). When scratched with a knife, or reduced to powder, it exhibits the red colour of hæmatite. The specific gravity of specular iron ore is 5·2. The island of Elba has long been noted for its specular iron ore, and it is also found in considerable quantity in Russia and Spain. The excellent iron of Nova Scotia is extracted from the specular ore of the Acadian mines.

Brown Hæmatite contains the same oxide of iron as red hæmatite, but in a state of chemical combination with water, the latter varying from ten to fourteen parts in a hundred parts of the ore. Its appearance varies widely in different specimens. Some form globular masses of considerable size, others occur in small round grains (*pea iron ore*); occasionally it occurs in stalactites; the various soft earthy *ochres** and *umbers*† consist, for the most part, of brown hæmatite, the dark brown shade, in the latter case, being caused by the presence of an oxide of manganese.

In this country brown hæmatite is found at Alston Moor (Cumberland) and in Durham, but it is a much more important iron ore in France. Some varieties of it contain a considerable proportion of phosphorus (in the form of phosphoric acid combined with oxide of iron), which materially affects the quality of the iron extracted from them.

The *Black Brush Ore* of the Forest of Dean is a brown hæmatite containing 89 parts of peroxide of iron, and 10 parts of water, and yields most of the iron used for making tin-plate.

The *Lake Ores* of Sweden consist of a variety of brown hæmatite ore which occurs at the bottom of lakes, and is collected by dredging with a kind of iron sieve attached to a long pole, and thrust down through a hole in the ice, the mud being washed away from the ore by shaking the sieve up and down under water. In this way a man will sometimes collect as much as a ton of ore in a day.

Spathic Iron Ore (*Spath* [German], spar), or *sparry iron ore*, is composed of carbonate of iron, or iron combined with oxygen and carbonic acid. Some specimens consist of a collection of nearly transparent shining crystals, which are almost colourless, and have the same crystalline form as calcareous spar (carbonate of lime). It is found in extensive beds in Styria and Carinthia. Spathic ore almost invariably contains manganese, which especially adapts it for the manufacture of certain kinds of steel, whence it has sometimes been termed *steel ore*. This ore is also found at Wearsdale in Durham, and on the Brendon Hills in Somersetshire.

Clay Iron Stone, *Argillaceous*‡ *Iron Ore* or *Clay-band*

* Derived from the Greek for *sallow*, in allusion to their yellow colour.

† From *umbra* (Latin), shade, on account of their darker tint.

‡ *Argilla* (Latin), clay.

contains the iron in the same form of chemical combination (carbonate of iron), in which it exists in spathic iron ore, but in a state of intimate mixture with clay. This ore is by far the most important of British iron ores, being that which is most extensively worked in this country. It occurs in great abundance around Dudley, in Staffordshire, in Yorkshire, Derbyshire, and in South Wales. Both this ore and the black-band ore are found in layers which occur alternately with beds of coal, limestone, clay, and shale, whence they are often spoken of as the *Ironstones of the Coal Measures*, and the circumstance that the same pit, or neighbouring pits, will furnish the coal employed for smelting the ore, the limestone used as a *flux*, and even the clay for making fire-bricks for the furnace, allows English iron to be produced at a price with which, until lately, other countries found it impossible to compete. The coal formations of Belgium and Silesia also furnish large supplies of clay ironstone.

This ore is found sometimes in continuous beds, and sometimes in irregular globular masses, imbedded in clay; it is moderately hard and stony, and varies in colour, through different shades of grey, slaty-blue, and brown. It is lighter than the preceding ores, and a cursory observer might regard it as a stone rather than a metallic ore. The proportions of carbonate of iron and clay contained in the ore vary considerably, the former amounting to 80 or 90 parts in the hundred in some specimens, and in others not exceeding half the weight of the ore.

Blackband Ore differs from clay iron-stone only by containing, in addition to its other constituents, a quantity of bituminous or coaly matter, sometimes amounting to one-fifth of the weight of the ore, and imparting the colour from which it derives its name. Lanarkshire and Ayrshire, in Scotland, contain extensive deposits of blackband iron ore, first brought to light in 1801. The presence of so much combustible matter allows the ore to be smelted with less expenditure of fuel. Blackband ores are also mined in Prussia.

The Northamptonshire iron ore, which is found in the *oolite* limestone in that county, appears to have been formed by the chemical alteration of clay iron-stone, under the influence of air and water, for it contains carbonate of iron associated with clay, and with the red oxide of iron, which is formed by the action of air upon the carbonate. Since the Northamptonshire ore contains much phosphorus (in the form of phosphoric acid), it yields an inferior iron.

*Iron Pyrites** is the yellow substance, of metallic appearance, which is so common in lumps of coal, and may be found in rusty globular masses on the sea-beach. It is composed of 46½ parts of iron in chemical combination with 53½ parts of sulphur, and is extensively used as a source of sulphur in the manufacture of sulphuric acid. Attempts have been made to extract iron from this mineral after the sulphur has been burnt off, and the iron is left in combination with oxygen from the air, but they have not been attended with much success. Since crystals of iron pyrites are found, not only in coal, but in many iron ores, it is the chief source of the sulphur which is the most objectionable impurity in iron.

EXTRACTION OF IRON FROM ITS ORES.

The difficulty of separating the iron from the other substances with which it is associated in the ore, is of course greater in proportion as these foreign matters are more numerous; thus, when the iron is combined with oxygen only, as in the magnetic iron ore, red hæmatite, and specular iron ore, the metal may be extracted at once in the form of malleable iron, by merely heating the ore in contact with carbon (charcoal), which combines with the oxygen;

* The word *pyrites* is derived from the Greek for *fire*, probably in allusion to the circumstance that when this mineral is heated it gives off sulphur, which takes fire.

and this, which is the primitive process for obtaining wrought iron, is still followed in places where the above ores can be readily obtained, and wood is abundant for conversion into charcoal. Even brown hæmatite and spathic ore can be treated in a similar manner if the water and the carbonic acid which they respectively contain be previously expelled by calcination.

But in the clay iron-stones and in blackband, the earthy matter (clay, &c.) which is present renders such a process impracticable, and it is necessary to raise such ores to a much higher temperature, in contact with lime, to liquefy the clay, so that it may be separated from the iron; when the high temperature causes the iron to combine with the carbon of the fuel, forming cast iron, from which the carbon is removed by a subsequent process, in order to obtain wrought iron.

Since clay iron-stones and blackband are the principal ores smelted in Great Britain, this is the process generally employed in this country.

Extraction of Iron from its Ores in the Form of Cast Iron. The ore to be smelted is broken up into lumps about twice the size of a fist, and in some cases it is found advantageous to prepare it for smelting by a preliminary process of *calcining* or *roasting*. For this purpose the ore is mixed with small coal, and built up, on a foundation of lumps of coal, into huge pyramidal heaps or *clamps*, which are kindled at the windward end, and allowed to smoulder for months, being prolonged, as may be requisite, by fresh additions of ore and fuel at the opposite extremity. The ore may be calcined with an addition of only one-twentieth of its weight of coal, if it contain, as is the case with blackband, a large proportion of bituminous or combustible matter, whilst a clay iron-stone may require as much as one-fifth of its weight of coal. This calcination in heaps is a very uncertain process, on account of the irregular distribution of the heat; some parts of the ore being scarcely affected, whilst others are over-heated and melted so that they can be smelted only with difficulty. This plan is only adopted in districts where fuel is very cheap.

During the process of calcination the ore loses about one-fourth in weight, in consequence of the expulsion of water and carbonic acid, and the combustion of the bituminous matter.

A portion of the sulphur from the pyrites contained in the ore is also burnt off during the roasting process, entering into combination with oxygen from the air, to form sulphurous acid gas.

Many ores are rendered much more porous by this process, and are then more readily smelted.

The expulsion of water and carbonic acid would, of course, be effected by the heat of the smelting furnace if this preliminary roasting were omitted; but the sulphur would not be removed to the same extent, and, since this is one of the most damaging impurities in iron, the roasting is a necessary preparation for ores containing much pyrites, and is always practised in Scotland.

During the calcination of the ore, the oxygen of the air combines with the protoxide of iron (ferrous oxide) which is contained in it, converting it into the sesquioxide (ferric oxide), which is less liable to enter into combination with the silica present in the ore, and thus to cause a loss of iron in the slag during the subsequent smelting process.

In South Wales the ore is roasted in furnaces, or *running-kilns*, resembling lime-kilns, into which it is thrown at the top, alternately with layers of small coal, the roasted ore being raked out at the bottom of the furnace.

Calcination has been much less practised since the introduction of the hot blast.

Process of smelting Iron Ores.*—Since this operation requires a very high temperature, it is carried out in a *blast-furnace*, which does not depend upon the draught of a

* From the German *schmelzen*, to melt.

chimney, but has a strong blast of air forced into it from beneath.

This blast-furnace varies much in form and dimensions,

Fig. 1.—Blast-furnace for Smelting Iron Ores.

according to circumstances, which will be better understood when the smelting process has been described. Fig. 1 shows a common form, and exhibits the essential features of its construction.

The chimney, or *tunnel-head* F, has three openings at the side, closed by iron doors, through which the ore and fuel are introduced into the furnace, when they fall through the *throat* or *tunnel-hole* G, into the *body* C, which is generally of a barrel-shaped form, widening as it descends, and thus allowing a free descent of the materials until they reach the *boshes* A, after which the furnace contracts rapidly, as at B, in order somewhat to check the descent of the solid materials, and attains its smallest diameter at the top of the *hearth* E, which is at its upper part a nearly cylindrical passage, but is made almost rectangular at the lower part, into which, through apertures *o o* in the walls, a blast of air is forced. The blast-pipes, or *tuyères*, are usually three in number, and are situated on three sides of the hearth. The fourth side is constructed differently from these three, its upper part being formed by a heavy block of stone *a* (Fig. 2), called the *tymp-stone*, which is supported by a cast-iron *tymp-plate* *b*, built into the masonry of the furnace, whilst the lower part is enclosed by the *dam-stone* *c*, faced externally with a thick cast-iron *dam-plate* *d*. That portion of the hearth which is shut in by the dam-stone is called the *crucible*, for it is here that the cast iron produced in the furnace accumulates, in a melted state, covered with a layer of melted earthy matter or *slag*.* The space between the tymp and dam stones is rammed up with good binding sand, in which an opening is made, just above the dam-stone, through which the slag

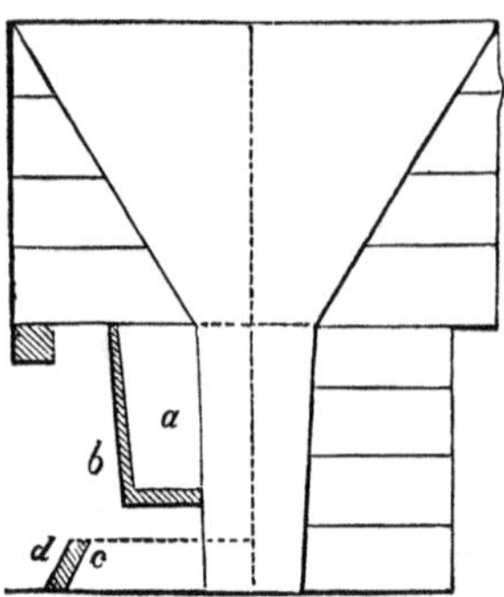

FIG. 2.—Boshes, Hearth, and Crucible of Blast-furnace.

* From the German *Schlacke*, dross.

is allowed to flow out of the furnace. The melted cast iron is never permitted to rise to the level of this opening, but is run out once or twice in twenty-four hours through a *tap-hole* at the bottom of the crucible, which is rammed up again with binding sand.

The dimensions of blast-furnaces vary much according to the conditions of their working, but some idea of them may be acquired from the following :—

Height of the blast-furnace, from 45 ft. to 100 ft.
Height of boshes from commencement of hearth, about 15 ft.
Greatest diameter at the boshes, from 13 ft. to 18 ft.
Diameter of chimney at charging-platform, about 10 ft.
Diameter of hearth, 3 ft. to 9 ft.
Depth of crucible, from 8 to 10 inches.
Total depth of hearth, 6 ft. to 8 ft.
Depth from tuyère holes to bottom of hearth, from 12 to 36 inches.

When the coal and ore are soft, and easily crushed, they are liable to obstruct the passage of the blast if there be too much pressure from the column of material above, so that the furnace must not be so high as when harder materials are employed. But in such a case, the diameter of the hearth and body of the furnace may be increased, whereas when hard ores and anthracite coal are employed, the diameter must be reduced, sometimes to one-half, and the height increased. The compression caused by the weight of materials in the furnace sometimes amounts to one-fourth of the bulk of the charge, so that 7,500 cubic feet of materials measured in the charging barrows may be thrown into a furnace of 6,000 cubic feet in capacity.

The tuyères,* or *twyers*, through which the blast of air is forced into the furnace, are formed of conical tubes of cast iron (*a b d c*, Fig. 3), having double walls between which water is introduced through the pipes *t t'*, and made to circulate as shown by the arrows, in order to keep them from being melted. They are between 2 and 3 inches wide

* From the French *tuyau*, a pipe.

at the opening into the furnace, and are built into the walls, as shown in Fig. 4, which represents a section of the hearth at the level of the tuyères. The openings are inclined somewhat in different directions, so that the blasts coming from them shall not exactly meet each other. The air is forced into the tuyère through a movable nozzle (B, Fig. 3) of copper or sheet-iron, connected by a leathern hose (A) with the pipe coming from the blowing machine, which is a large forcing pump worked by steam power, and capable of supplying air, at the rate of from 4,000 to 10,000 cubic feet per minute, to each furnace.

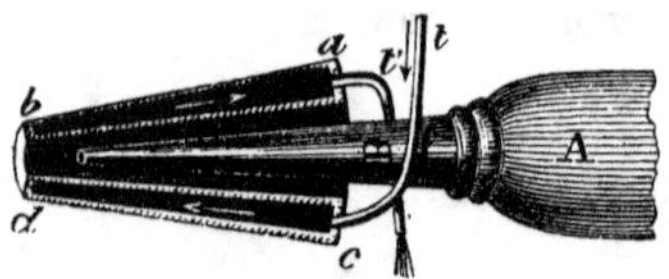

FIG. 3.—Tuyère of Blast-furnace.

The pressure of the blast as it issues into the furnace usually amounts to $2\frac{1}{2}$ lbs. or 3 lbs. upon the inch, but it is sometimes increased to 5 lbs. In charcoal furnaces, the fuel being very porous, a pressure of 1 lb. upon the inch is sufficient; in furnaces fed with coke, about 3 lbs. upon the inch; but where anthracite is employed, the blast has a pressure of 4 lbs.

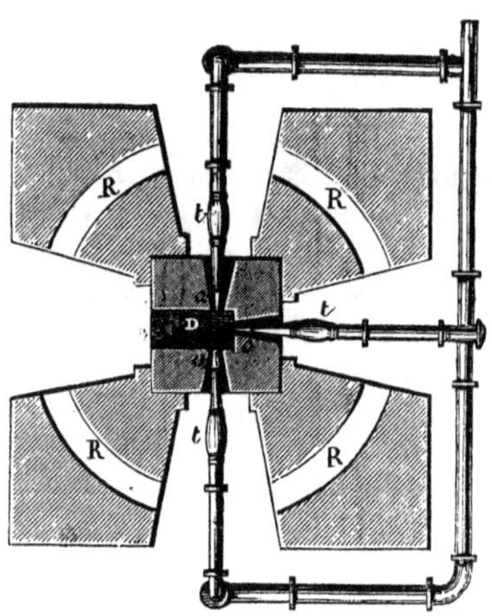

FIG. 4.—Arrangement of Tuyères in Blast-furnace. *t*, Pipes conveying the blast. *o*, Openings in the sides of the hearth D. R, Arched passages for the labourers.

When a blast of high pressure is employed, it becomes

necessary to stop up the space between the tuyère and the nozzle of the pipe conveying the blast.

The elbow-pipe, which connects the blast-pipe with the air main (Fig. 4), has usually a small peep-hole closed with glass or mica, so that the temperature of the hearth, as indicated by the colour of its glow, may be observed by looking through the tuyère.

In some modern furnaces the number of tuyères has been much increased. *Rachette's blast-furnace*, which is used in the iron-works of the Ural for smelting magnetic ore, and yields 30 tons of iron in 24 hours, is a furnace of a narrow oblong section, with eight tuyères on each side. It widens as it ascends, unlike the older blast furnace, so that the width is greatest at the mouth, which causes the charge to sink more uniformly in horizontal layers.

When the air is blown into the furnace at the ordinary temperature of the atmosphere, it is spoken of as a *cold-blast* furnace, whilst a *hot-blast* furnace is one in which the air is heated to 500° or 600° F. (about the melting point of lead), by being forced through a considerable length of red-hot iron pipes, before it enters the furnace.

At first sight it would appear to be immaterial whether the blast of air be heated before or after it is forced into the fire, since, in either case, the same quantity of heat must be lost in raising the air to the temperature of the fire itself, whether a part of this heat be produced by fuel burnt in order to warm the iron pipes, or the whole of it be produced by fuel consumed in the furnace. In the stoves for heating the blast, however, the carbon of the fuel is converted into carbonic *acid*, whilst in the blast-furnace it is converted into carbonic *oxide*, combining, in the latter case, with only half as much oxygen as in the former, and producing less than one-third as much heat, so that 2 cwt. of coal burnt in the stove will go as far as above 7 cwt. burnt in the furnace. Moreover, when cold air is blown into the furnace, it must pass through a much larger quantity of the heated fuel before it is raised to the temperature of the fire than would be the case if it had been heated before entering, so that the hot blast has less cooling effect upon the fire, and a much higher temperature may be produced by the same consumption of fuel, or a fire sufficiently hot for smelting the iron ore may be raised with a smaller consumption of fuel than when a cold blast is employed. The economy of fuel is still greater when the blast is heated without any extra expenditure of fuel, by employing the waste heat derived from the furnace itself.

An improved method of heating the blast consists in employing two chambers of fire-brick, in which fire-bricks are loosely stacked. Into each of these chambers, alternately, the flame of a coal fire is passed until the bricks are heated to redness; the fire is then diverted into the other chamber, whilst the blast is sent over the hot bricks, when it becomes heated to 1,300° F. before entering the blast-furnace. By the time this chamber has been cooled, the other is heated and ready to do duty again. These *regenerative stoves*, as they are called, have been employed, with great advantage, in other cases.

In the construction of the blast-furnace, very infusible materials are necessary, especially in the crucible and hearth, where the highest temperature prevails. For these portions of the furnace, millstone grit or Newcastle sandstone is sometimes employed (though fire-brick has been lately used), but the upper part or body of the furnace, which is not exposed to so high a temperature, is lined either with fire-brick, or sometimes with blocks of *slag* from the furnace itself, which undergoes an alteration in structure, rendering it less fusible, when exposed to the heat of this portion of the fire. The fire-brick lining (*i l*, Fig. 1) of the furnace is double, a space of about three inches being left between the two portions, and rammed up with powdered coke or sand, which yields to the expansion of the brick-work, and hinders the conduction of the heat to the external part of the structure. Much attention is given to the slope of the boshes, for if this be too steep, the materials will fall too quickly, and if it be not steep enough, they will stick to the sides, and form obstructions or *scaffolds*, preventing the descent of the materials above. Three or more blast-furnaces are generally built in a row, and vaulted passages are left around the hearth to allow easy access to the tuyères and the crucible. When the masonry of the blast-furnace is slight, it is strengthened by iron hoops and bars (*cupola* furnace), or it is sometimes cased externally with boiler-plates riveted together.

The *fuel* of blast-furnaces in this country is almost exclusively coal or coke, for charcoal is far too expensive for general use, though its freedom from sulphur enables a better quality of iron to be manufactured with it. In Austria, charcoal from wood or peat is the only fuel employed, and iron of the finest quality is the result. The Swedish iron works also use charcoal only.

In hot-blast furnaces, coal may be employed without having been coked, because the higher temperature which prevails in the furnace allows it to become converted into coke in the upper part of the fire; but for cold-blast furnaces, coke is the ordinary fuel, a hard-burnt dense quality being preferred.

For lighting or *putting the furnace in blast*, much time is required, since there would be great danger of cracking the lining by a sudden application of a very high temperature. Before the dam-stone is fixed in its place, faggots are kindled in the opening below the tymp-stone, so that the flame and heated air may be drawn up into the furnace and gradually warm it. After a few days, a quantity of coal or coke is thrown in at the throat of the furnace, and gradually increased until the furnace is filled; the blast is then gradually turned on, and when the fuel has sunk down sufficiently, the smelting operation is commenced. This drying and warming of the furnace occupy a month or more, according to its size.

For the convenience of charging the furnace with fuel and ore, a gallery runs round it at the level of the tunnel-hole, to which the materials are brought in iron waggons or barrows, either up an inclined plane, or along a tramway carried from the slope over against the furnace.

From 3 to 6 cwt. of iron ore, according to its richness, is thrown on to the top of the burning fuel, together with about one-third of its weight of limestone or of quick lime, which is employed as a *flux**, to bring the clay of the ore to a liquid state in the fire, in order that it may be separated from the iron. Upon this a charge of fuel is thrown, consisting of about 4 or 6 cwt. The descent of the charge in the furnace is very gradual, from 40 to 50 hours being usually occupied in its passage from the top to the bottom.

After an interval of half an hour or so, fresh charges of ore, flux, and fuel are introduced, and the furnace is thus fed, night and day, for six or seven years, before it is found necessary to *blow it out* in order that its lining may be repaired.

The melted cast iron which is produced runs down into the crucible, together with the liquid slag, formed by the action of the lime in the flux upon the clay in the ore. The cast iron, being the heavier of the two, accumulates at the bottom, and above it, five or six times its bulk of liquid slag, which flows out over a notch in the dam-plate, and is generally received in iron moulds and cast into large blocks of 15 or 20 cwt. each.

At intervals of twelve hours, generally, the furnace is *tapped* by opening the tap-hole with an iron rod, and a *cast* is made, that is, the liquid cast iron is run out and cast into *pigs*, or rough half-round bars, either in thick iron moulds, or in trenches excavated in strongly binding sand, and communicating with a central channel into which the cast iron flows from the furnace. The pigs of iron are about three

* From the Latin *fluo*, to flow.

feet long and four inches in diameter, and weigh $2\frac{1}{2}$ cwt. About five or six tons are usually run out at each tapping.

The *foundry iron* intended for the manufacture of castings

FIG. 5.—Hot-Blast Furnace for Smelting Iron Ores; showing, on the right, the stove for heating the blast; on the left, the large air-chamber or regulator for equalising the blast. In front are seen the slag-moulds on a tramway, and the channel through which the cast iron is conducted into the pig-moulds.

is generally run into sand-moulds; but *forge iron*, which is to be afterwards converted into wrought iron, is cast in iron moulds, partly to avoid contamination with the sand, and partly to render the metal brittle, so that it may be easily broken up for the puddling process which it next undergoes.

The blast is suspended during the operation of tapping, and turned on again at the end, in order to force all the iron and slag out of the hearth. Fig. 5 exhibits a general view of the arrangements connected with a blast-furnace.

Chemical Changes which take place in the Blast-furnace.—The great variety of the substances present in the blast-furnace causes the chemical reactions to be too numerous and elaborate to be fully considered except in a purely chemical treatise; but the principal changes which result in the production of cast iron and slag from the ore, flux and fuel can be easily understood.

The iron being contained in the ore in the state of oxide, that is, in chemical combination with oxygen, this element must be removed in order to obtain the iron in the metallic state. The whole of the iron would thus be easily *reduced* by the combustible gases in the furnace, if the ore consisted of oxide of iron only, but the presence of clay in the iron-stone introduces a difficulty. Clay consists of alumina and silica, and cannot be melted alone, even in the blast-furnace. In contact with oxide of iron, however, the clay combines chemically with that substance, and melts to a liquid which becomes a black glass or slag when it cools. The greater part of the iron would thus be lost in the slag. But if a quantity of lime be mixed with the ore, the clay will combine with it instead of combining with the oxide of iron, which can then be made to yield the whole of its metal. The lime is sometimes put into the furnace in the form of limestone or carbonate of lime, and sometimes in that of quicklime, or limestone from which the carbonic acid has been expelled by heat. The use of quicklime not only avoids the local loss of heat necessary to expel the carbonic acid from the limestone, but it increases the production of iron, because it occupies a smaller space than the limestone, and allows more charges to pass through the furnace in a given time.

The air which is blown in at the bottom of the furnace contains one part by measure of oxygen mixed with four measures of nitrogen. When it comes into contact with the glowing carbon of the fuel, the latter combines chemically with the oxygen, forming carbonic acid gas, which contains eight parts by weight of oxygen combined with three parts of carbon; but as this gas passes up the furnace, it gives up one half of its oxygen to another portion of the heated carbon, and becomes carbonic *oxide* gas, which contains four parts by weight of oxygen combined with three parts of carbon. Still passing upwards through the furnace, the carbonic oxide meets the red-hot ore containing oxide of iron, from which it regains the oxygen necessary to convert it into carbonic *acid*, thus reducing the iron to the metallic state. Continuing its ascent, this carbonic *acid* traverses more red-hot carbon, and is again converted into carbonic *oxide*, in which form it issues from the chimney of the furnace, accompanied by the nitrogen of the air and by smaller quantities of other gases produced in the furnace.

The metallic iron, which has been separated from the oxygen during the passage of the carbonic oxide over the heated ore, is not capable of being melted at the temperature of this furnace until it has combined chemically with some of the carbon which surrounds it, and has thus become converted into *cast iron*, which is then liquefied by the heat and runs down into the crucible of the furnace, accompanied by the melted slag formed by the combination of the lime in the flux with the clay contained in the ore.

The chief chemical operations taking place in the blast-furnace may therefore be summed up as follows :—

1. Combination of oxygen from the air with carbon from the fuel to form carbonic *acid*.
2. Conversion of carbonic *acid* into carbonic *oxide* by giving up half its oxygen to red-hot carbon.
3. Conversion of oxide of iron into metallic iron by parting with its oxygen to heated carbonic *oxide*.

4. Combination of metallic iron with carbon to form cast iron.
5. Combination of lime with clay (silica and alumina) to form a glass or slag.

Employment of the waste Gas from the Blast-furnace.—The gas which issues from the chimney of the furnace contains, beside the carbonic oxide, considerable quantities of other combustible gases produced by the action of heat upon the coal when that fuel is employed, and of hydrogen derived from the decomposition of the water present in the air and in the solid materials. This gas was formerly allowed to burn in the air as it escaped from the chimney, and its flame, coloured yellow and red by minute quantities of vapour of sodium and calcium, illuminated the sky in the iron districts. But within the last thirty years the waste gases have been turned to more useful account. Openings *aa* are sometimes made in the sides of the furnace A, as shown in Fig. 6, at about four feet below its mouth, through which the gases find an easier escape than through the mass of fuel and ore above, and pass into flues, whence they are conducted by iron pipes *b* into kilns, where they are used as fuel for calcining the ore or the limestone flux.

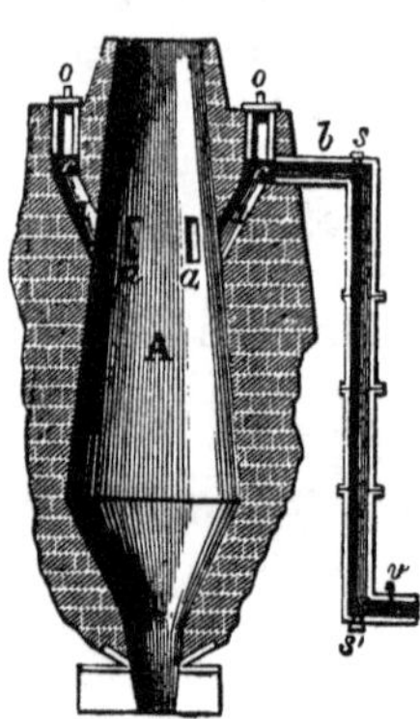

FIG. 6.—Blast-furnace, with openings for drawing off the waste gases.

In order to allow the waste gas to be drawn off more completely, the mouth of the furnace is sometimes provided with a hopper, as shown in Fig. 7, the lower part of which is closed by an inverted cone of sheet iron, suspended by a chain from one end of a lever and balanced by a weight at the other. When the charge is thrown into the hopper the cone descends and the materials fall into the furnace, after which the cone returns to its former position, closing the hopper, and allowing the waste gas to pass off through the flue.

At Ulverstone the waste gases are drawn off through an iron pipe about three feet wide, which descends six feet into the furnace, and is supported by brickwork projections. (Fig. 8.) Exhausting fans are employed to draw the gases through the furnaces in which their combustion is effected. The pipe, being placed in the centre of the throat of the

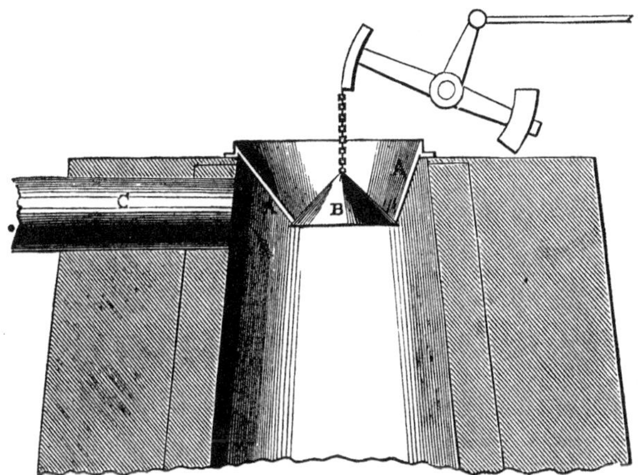

FIG. 7.—Cup and Cone for closing the Blast-furnace, in order that the waste gases may pass into the lateral flue, as shown by the arrow.

furnace, leaves an interval of about four feet all round it, for the introduction of the charge, which is thus equally distributed.

The large proportion of carbonic oxide present in the gas escaping from the blast-furnace, renders it very poisonous, and it has occasionally caused the death of persons engaged in tending the furnace.

Although by far the greater part of the nitrogen contained in the air blown into the furnace undergoes no chemical change, but escapes unaltered in the waste gases, a small proportion of it appears to enter into combination with carbon, and with the potassium which is present in the materials composing the charge, to form the cyanide of potassium, which has been obtained in considerable quantity from some furnaces. Beautiful copper-coloured crystals are not uncommonly found in the hearth of the furnace, consisting of a compound of nitrogen with carbon and titanium, this metal being derived from the titanic acid which is frequently present in the ores of iron.

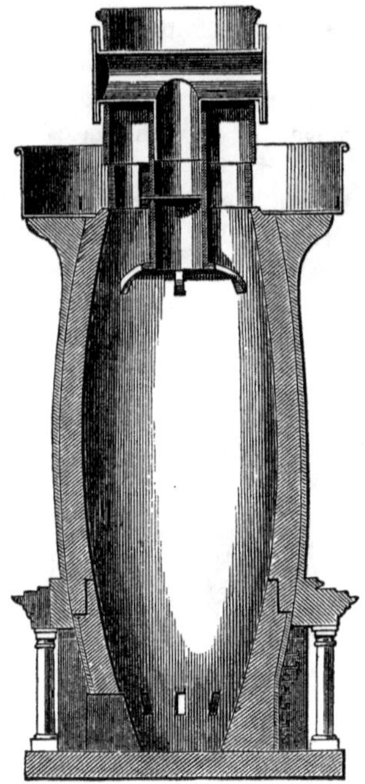

FIG. 8.—Schneider's arrangement for collecting the waste gases from the Blast-furnace by a central pipe.

As might be expected, the composition of the gas issuing from the furnace varies considerably; but the subjoined analysis of a sample, taken from a furnace in which coal was the fuel employed, will illustrate its general nature.

Gas from Blast Furnace.

Nitrogen	55 vols.
Carbonic Oxide	26 ,,
Hydrogen	7 ,,
Carbonic Acid	8 ,,
Marsh Gas	3½ ,,
Olefiant Gas	0½ ,,
	100

Composition of the Slag or Cinder from the Blast-furnace.—The slag which runs out of the furnace solidifies on cooling to an opaque glassy mass, in various shades of grey, blue, green, brown, and black, sometimes prettily striped and variegated. It consists essentially, as already noticed, of the constituents of clay, viz. silica and alumina, united with lime, and the presence of these three substances in the furnace is necessary for the production of a liquid slag. In order to secure this, both the quality and quantity of the earthy matter or *gangue* * associated with the ore must be taken into account. When clay, which contains both silica and alumina, is present, lime will form an appropriate flux; but if limestone only were contained in the ore, an addition of clay would be requisite; again, ores in which the gangue consisted of quartz (silica) would require the addition of alumina (in the form of clay) as well as lime. In some cases it is found possible to mix the different ores so that the clay associated with one may form a slag with the lime contained in another, rendering it unnecessary to add any other flux.

In order that clay and lime may easily melt together into a glass, the lime should be rather more than one-third of the weight of the clay; and since limestone (or carbonate of lime) contains somewhat more than half its weight of lime, two parts of limestone should be employed for three of clay. In furnaces where coal or coke is burnt, it is usual to employ rather more lime than this, because it hinders the sulphur from passing into the iron; for lime consists of the metal calcium in chemical combination with oxygen, which is removed by the strongly-heated carbon, leaving the calcium free to unite with the sulphur, forming a sulphide or sulphuret of calcium, which is found in the slag. It is said that not only more sulphur but also more silicon is found in the iron when less limestone is employed. An average

* From the German *Gang*, a mineral vein.

sample of slag from the blast-furnace gave the following results when analysed:—

Slag from Blast Furnace.

Silica	43·07
Alumina	14·85
Lime	28·92
Magnesia	5·87
Oxide of Iron	2·53
Oxide of Manganese	1·37
Potash	1·84
Sulphide of Calcium	1·90
Phosphoric Acid	trace
	100·35

The slag is commonly employed for road-making in the neighbourhood of the iron-works. Some attempts have been made to turn the slag to account by employing it as a manure for soils deficient in potash, of which it will be seen that the above slag contains nearly $\frac{1}{50}$th of its weight, in a form which would be easily rendered available for plants by the combined action of air and moisture. When the slag is run into water, or blown into a frothy condition by the blast, it resembles pumice-stone, and is easily ground to a powder fit for applying to the soil.

It might be anticipated that the appearance of the slag would convey to the experienced eye some useful information with respect to the character of the ore and the general progress of the smelting operation. A good slag is liquid, nearly transparent, of a light grey colour, and has a fracture somewhat resembling that of limestone. A dark slag shows that much of the oxide of iron is escaping unreduced. Streaks of blue are commonly found when ores containing sulphur are being smelted, possibly from the presence of a substance similar to ultramarine, the constituents of which are all present in the slag. Again, the slags obtained in smelting ores containing titanium generally present a peculiar blistered appearance. The connection between the character of the slag and that of the cast iron obtained at the same time will be particularly noticed hereafter.

Some of the slag usually accumulates in front of the nozzles of the tuyères, where it is chilled by the blast, and forms a conical prolongation of the pipe, which is useful in directing the blast into the centre of the furnace instead of allowing it to pass up the sides. But occasionally these *noses* of slag meet in the middle and obstruct the blast in a very serious manner. The proportion between the iron and the slag is of some importance; for if the volume of the latter be too small, it will not suffice to protect the metal from the action of the oxygen in the blast, and much waste of iron, in the form of oxide, will result.

The slag should require about the same temperature to melt it as the cast iron; for if it be too fusible, a part of the oxide of iron contained in the ore may be dissolved in it and escape reduction to the metallic state.

CAST IRON.

The only substance with which the iron is invariably and indispensably associated in cast iron is carbon, so that it has sometimes been spoken of as a *carbide* or *carburet* of iron. But this would convey the impression that the whole of the iron present existed in chemical combination with the carbon, exactly as, in the *oxide* of iron, the whole of the iron is combined with oxygen, whereas the greater part of the iron in cast iron is present in the uncombined state, its properties being altered or modified by the presence of the carbide of iron (or compound of carbon with iron) which is diffused throughout the metal. But for the circumstance that the term *alloy* is conventionally restricted to the substances formed by the alliance of *metals* with each other, whilst carbon is a non-metallic body, cast iron might be appropriately designated an alloy of iron with carbon.

By fusing finely divided iron with charcoal until the metal has taken up as much carbon as it will dissolve, a dark grey mass is obtained, which is so brittle that it may be powdered in a mortar. This substance appears to be a chemical compound of carbon with iron, or a carbide (carburet) of iron, containing, in 100 parts, 94·36 of iron, and 5·64 parts of carbon. The proportion of carbon present in cast iron varies in different samples, but it never amounts to 5 per cent., so that the carbide of iron must be regarded as being in a state of intimate mixture with metallic iron.

On examining the fracture of freshly-broken pieces of cast iron, it will be found that some specimens have a silvery white, and others a grey colour, caused by the presence of very minute particles of carbon, which are interspersed among the lighter-coloured particles of the metal. When the grey samples of cast iron are acted upon by acids (diluted sulphuric or hydrochloric) the iron is dissolved, but the black particles of carbon are left, and these are found to possess the same properties as the natural variety of carbon known as black lead or *graphite*,* of which pencils are made. When the white cast iron is dissolved in acids, very little black residue of carbon is left, because the greater part of the carbon is present in the state of chemical combination with the iron, as a carbide of iron which is dissolved by the acid, and very little is present in the form of graphite.

When a sample of grey cast iron is melted, the particles of free carbon are dissolved by the liquid metal, becoming chemically combined with a portion of the iron; and if the melted mass be suddenly chilled by throwing water upon it, or by running it, when not far from its point of solidification, into a thick iron mould, the carbon does not separate again, so that a mass of white cast iron is thus produced. It is more difficult to convert the white into the grey variety of cast iron, but this has been effected by exposing the melted metal to a high temperature and allowing it to cool down very slowly, when a portion of the carbon separated from the iron and the grey variety of cast iron was produced.

* From the Greek verb, *to write.*

These principles receive important practical application in the process of *chill-casting*, in which the iron is cast in thick iron moulds or *chills*, in order to impart to it a hardness rivalling that of steel. On the other hand, castings which are too hard to admit of being turned or bored, are softened by being heated for several hours in sand, or in a mixture of coal-dust and bone-ash, being afterwards allowed to cool slowly, which is favoured by the bad conducting power of those materials. When it is required to produce a casting of which one part is white and hard, whilst the other is grey and tough, the former is cast in a chill and the latter in sand. Thus, plough-shares are cast in a mould of which one side is made of thick cast-iron, and the other of moulder's sand, so that the grey iron formed in the latter strengthens the hard chilled white iron. In a similar manner, the heads of shot or shell intended for penetration are cast in chills, but the bodies which need not be so hard, and should be of a tougher metal, are cast in sand.

Since in grey cast iron a smaller proportion of the iron is in combination with carbon, and more of it in the true metallic state, this variety would be expected to exhibit more of the properties of metallic iron, whilst white iron ought to present the characters of the chemical compound of carbon with iron, described above. Accordingly, the grey cast iron is much softer than white iron, and admits of being filed and turned, whereas the white iron is so extremely hard that a file will scarcely touch it, and a blow from a hammer, which indents a pig of grey iron, will break up one of white iron. The larger proportion of metallic iron contained in the grey cast iron causes it to require a higher degree of heat before it begins to exhibit signs of fusion, but it is capable of becoming very liquid at a sufficiently high temperature, so as to be easily run into moulds; white cast iron, on the other hand, is softened at a rather lower temperature, but does not flow well, assuming a somewhat viscid or pasty consistence. It *scintillates*, or throws off sparks or *jumpers*, as it runs from the furnace, to a much greater extent than grey iron. White cast iron is about $\frac{1}{20}$th heavier than the grey variety, its average specific gravity being 7·5, whilst that of grey iron is 7·1. The grey iron rusts more easily in air, and is more readily acted on by acids, than white iron, which may be ascribed partly to its containing more iron in an uncombined form, and partly to the acceleration of chemical action caused by the voltaic disturbance excited by the contact of the particles of graphite with the particles of iron, in the presence of the acid (in the case of air, carbonic acid).

White iron usually, but by no means invariably, contains less total carbon than grey iron.

Mottled cast iron is composed of a mixture of the grey and white varieties in varying proportions, the grey iron sometimes appearing in specks, like minute flowers, upon a white ground, whilst in other specimens the mass is composed of grey iron, and the white iron appears in spots.

The tenacity of cast iron is found to be increased by remelting, probably because some of the uncombined carbon is thus induced to combine with the iron.

Although, as stated above, carbon appears to be the only substance indispensably associated with the metal in cast iron, the commercial varieties of this material always contain silicon, phosphorus, sulphur and manganese, which are often present in considerable proportion, and are known to exercise an influence upon the character of the cast iron. Other substances, such as titanium, cobalt, nickel, chromium, copper, vanadium, calcium, magnesium and arsenic, may also be discovered by a careful analysis of considerable quantities of cast iron, but they are generally present in very small proportion, and are not known to produce any effect upon the metal.

Next to the carbon, *silicon* (or *silicium*) is the commonest and most abundant constituent of cast iron, its quantity varying from $\frac{1}{1200}$th part to nearly $\frac{1}{20}$th part of the weight of the metal, the proportion of silicon being higher in the grey than in the white variety. The silicon appears to exist in a state of chemical combination with the iron, and is derived from the silica in the ore or in the fuel. Silica is a combination of silicon with oxygen, and when the latter is abstracted by the carbon at the high temperature of the blast-furnace, the silicon enters into combination with the iron. The presence of a large proportion of silicon in cast iron is generally regarded as injurious to its quality, the strongest cast irons being those which contain a small quantity of that element. Iron which has been smelted with coke contains a larger proportion of silicon than that smelted with charcoal, and hot-blast iron commonly contains more than that smelted by cold-blast.

Manganese is seldom if ever absent from cast iron, for it is a metal which very nearly resembles iron in its chemical properties, and is commonly found in iron ores, so that the same operation which reduces the iron in the blast-furnace also reduces the manganese, and this metal becomes alloyed or intimately mixed with the melted iron. The manganese has been found in the large proportion of $\frac{1}{16}$th of the weight of the cast iron, but it seldom exceeds $\frac{1}{40}$th. The influence exerted by manganese upon the character of the cast iron is very decided, tending to the production of the white variety, the manganese diminishing the tendency of the carbon to separate in the form of graphite. White cast iron, therefore, is commonly found to contain the largest proportion of manganese.

The spathic iron ores yield a cast iron containing a particularly large quantity of manganese, sometimes exceeding $\frac{1}{10}$th of the weight of the cast iron. Such an iron is capable of retaining upwards of $\frac{1}{25}$th its weight of carbon in chemical combination with it, and the compound thus formed crystallises in large shining plates, whence it is named by the Germans *Spiegel-eisen* or *mirror-iron*. It is largely employed in the manufacture of Bessemer steel.

It has been asserted that the presence of manganese in

iron ores encourages the passage of sulphur and silicon into the slag, thus reducing the proportion of those injurious impurities in the metal.

Phosphorus occasionally forms between $\frac{1}{50}$th and $\frac{1}{60}$th part of the weight of cast iron, but about $\frac{1}{100}$th part is a commoner proportion of phosphorus. It exists in chemical combination with a portion of the metal, as *phosphide* of iron, and is derived either from phosphate of iron contained in the ore, or from phosphate of lime which is frequently present in the limestone employed as a flux, and in minute quantity in the coal. These phosphates contain phosphorus in a state of combination with oxygen, which is abstracted by the carbon of the fuel in the blast-furnace, and the phosphorus thus set free enters into combination with the iron. So completely is the phosphorus taken up by the metal, that only traces of that element, in the form of phosphates, are usually found in the slag from the blast-furnace. The effects of phosphorus are to harden the cast iron and to increase its fusibility.

Sulphur, though almost invariably contained in cast iron, rarely forms as much as $\frac{1}{50}$th of its weight. It may be derived, as already stated, from iron pyrites contained in the ore or in the coal. The white varieties of cast iron contain a larger proportion of sulphur than the grey, and it is generally admitted that the presence of sulphur diminishes the strength of cast iron in a very high degree.

The following table illustrates the general composition of the three principal varieties of cast iron :—

	Grey	Mottled	White
Iron	90·24	89·31	89·86
Combined Carbon .	1·02	1·79	2·46
Graphite	2·64	1·11	0·87
Silicon	3·06	2·17	1·12
Sulphur	1·14	1·48	2·52
Phosphorus	0·93	1·17	0·91
Manganese	0·83	1·60	2·72
	99·86	98·63	100·46

It must be observed that the difficulties attending the chemical analysis of cast iron are very great, on account of the large quantity of iron which has to be separated from small quantities of the other constituents, so that, although numerous analyses are recorded, their results do not exhibit that agreement which is necessary in order that the composition of this material may be considered to be thoroughly established.

The average tensile strength of cast iron is about seven tons per square inch.

For the useful applications of cast iron, eight varieties are commonly recognised. Nos. 1, 2 and 3 are decidedly grey irons, of different shades, 1 being the greyest; they are distinguished by the sparkling largely crystalline appearance of the broken surface, and are called *melting iron* because they are chiefly used for fine castings. No. 4 is *best grey forge* iron, and No. 5, *grey forge*; they do not become so liquid when melted as the preceding, but they are tougher and better fitted for purposes where strength is required. No. 5 is also used for the manufacture of wrought iron, as is also No. 6 or *strong forge*, which is still less grey in quality. No. 7 is a decidedly mottled iron, and No. 8 is white.

A less elaborate classification is generally employed among engineers; Nos. 1, 2 and 3 being made to include all shades, from dark grey (No. 1) to white (No. 3).

Grey cast iron is usually regarded as the proper or normal product of a blast-furnace in good working order, supplied with a due proportion of fuel, and the separation of the shining scales of graphite or *kish* upon the surface of the cast is commonly looked for as an indication of the satisfactory performance of the furnace. If the quantity of fuel employed be too small in proportion to the ore, so that the temperature is low, a white cast iron is produced, and a larger proportion of iron is found in the slag, to which it imparts a nearly black colour. The slag from a furnace yielding white iron frequently contains about 5 per cent. of the metal, whilst the slag from grey iron seldom contains more than 2 per cent. Slags which contain much iron, like those from white iron, also contain appreciable quantities of phosphorus in the form of phosphoric acid. When the proportion of sulphur in the fuel is very large, it is difficult to produce anything but white iron, except with a very large proportion of lime in the charge (see page 35).

At the Dowlais Foundry the capacity of the furnaces is 275 cubic yards. When grey iron is manufactured, these are charged, *for every ton of iron obtained*, with

48 cwt. Calcined Ore (Clay Ironstone)
50 ,, Coal
17 ,, Limestone

the blast being supplied at the rate of 5,390 cubic feet per minute.

When white forge iron, for conversion into bar, is made, the same furnaces are charged, *for every ton of iron*, with

28 cwt. Calcined Ore
10 ,, Hæmatite
10 ,, Slag (Forge and Finery Cinders)
42 ,, Coal
14 ,, Limestone

the blast being supplied at the rate of 7,370 feet per minute. In this case, then, one-sixth *less coal* and one-third *more air* are employed for the production of the white than for the grey iron. The weekly produce of the furnace is 170 tons when white iron is being made, and only 130 tons of grey iron. The slag is about twice the weight of the metal. The production of iron in winter exceeds that in summer by four or five per cent., because, in the latter season, there is more vapour of water in the blast, which lowers the temperature of the furnace.

The position and direction of the tuyères materially influence the character of the pig iron obtained in the crucible; when *forge pig* is being made, they are more inclined downwards so as to direct the blast upon the metal and partially refine it, but when *foundry iron* is the desired product, they are raised higher above the bottom stone of the furnace, so that the blast shall not oxidise the cast iron in the crucible. The influence of a large relative volume of slag in favouring the production of grey iron also depends upon its shielding the metal from the oxidising influence of the blast.

The production of white iron when the supply of fuel is deficient, has been plausibly referred to the lower temperature of the furnace, on the supposition that the particular compound of carbon with iron, which is present in white iron, is decomposed by a higher temperature (when more fuel is employed) into free carbon, which separates as graphite on cooling, and a compound of iron with the smaller proportion of carbon which exists combined in grey iron. The tendency of manganese to favour the production of white iron may be due to its forming a compound with carbon which resists decomposition at a high temperature, and the chemical similarity existing between manganese and iron favours the belief that a carbide of manganese dissolved in iron would produce an influence similar to that of a carbide of iron upon the properties of the metal.

It sometimes happens that grey and white cast iron run from the crucible of the furnace at the same tapping, in consequence of some variation in the progress of the smelting operation during the interval which has elapsed since the crucible was last tapped. Advantage is sometimes taken of this, in order to form mottled iron for casting rollers, by first conducting the process so as to obtain grey iron, and then increasing the charges of ore, decreasing the fuel, and applying a higher pressure of blast, so as to produce a white iron.

The belief is very common that iron smelted with the hot-blast is inferior in quality to that obtained from a cold-blast furnace, though it does not clearly appear to have been proved that, all other conditions being the same, the temperature of the blast does make a difference in the quality of the cast iron obtained. The smaller quantity of fuel employed in a hot-blast furnace might be the cause of the alleged inferiority of the iron. The higher temperature in the hearth of the hot-blast furnace also facilitates the formation of a pig iron rich in silicon. Moreover, part of the charge introduced into a hot-blast furnace often consists of slag or *cinder* formed during the process of puddling the cast iron in order to convert it into bar iron. This cinder contains more than half its weight of iron which it is desired to recover, but it also contains a large proportion of phosphorus and sulphur, by which the quality of the cast iron is deteriorated. It is not uncommon to employ a mixture of ore with ¼th of this cinder in a hot-blast furnace, and the *cinder iron* so obtained is inferior in quality and lower in price than the *mine iron*, which is smelted from ore only. The temperature prevailing in a cold-blast furnace is not usually adequate to the smelting of such slags, so that cold-blast iron is not so liable to impurities from this source.

Some of the latest experiments upon the comparative strengths of hot-blast and cold-blast irons appear to warrant the conclusion that so far as the temperature of the blast only is concerned, the hot-blast tends slightly to injure the quality of the softer (grey) irons, whilst it improves, sometimes in a very marked degree, the character of the harder (white) cast irons.

Remelting of Cast Iron in the Foundry.—In order to melt the pig iron for the production of castings, a *cupola furnace* is employed. The construction of this furnace varies in different foundries, but it is now commonly made of the form represented in Fig. 9, being cased with thick iron plates strongly riveted together, and protected internally by a layer of binding sand about nine inches thick. At different heights up the sides of the furnace there are openings G for the introduction of blast-pipes, two or three of which are employed at one time, the remaining openings being closed by iron doors. For melting 5 tons of cast iron, the cupola furnace is 9 feet high, and 3½ feet in diameter, clear of the lining. The tuyère holes, 6 inches in diameter. are placed at intervals of 15 inches, the lowest being 30 inches from the bottom, which is made to slope a little towards the gutter E by which the melted iron is run out. A conical iron hood D surmounts the furnace, connecting it with the chimney, and having an opening for introducing the metal and fuel. A wood fire is first lighted in the cupola, through the tap-hole, and a quantity of coke having been thrown in, and well ignited by the blast, the tap-hole is closed, and the pig iron introduced in pieces of about 28 lbs. each, together with one-fourth of its weight of coke. Fresh charges are introduced about every quarter of an hour, until the whole of the iron is melted, when the tap-hole is unstopped and the metal run, either into moulds sunk into the floor upon which the furnace stands, or into a casting-ladle, whence it is transferred to the moulds.

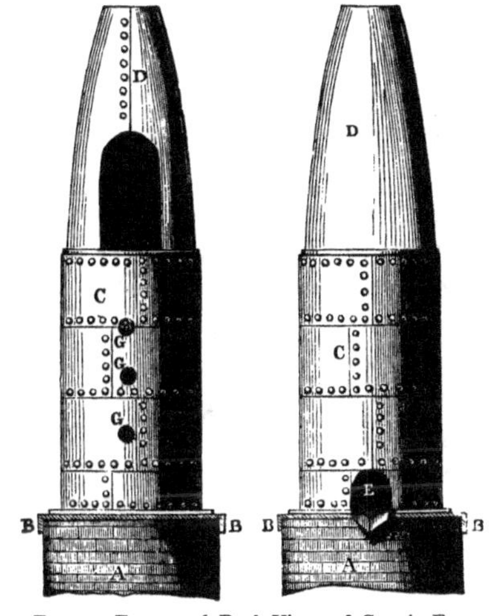

FIG. 9.—Front and Back Views of Cupola Furnace. A, Foundation of Masonry covered with Iron Plates B.

During the process of remelting, about five or six parts of iron out of a hundred are wasted by combining with the oxygen of the air to form an oxide of iron, which unites with silica from the lining of the furnace to form a slag. The quality of the iron is generally improved in the process in consequence of the refining influence of the air, and a highly carburetted iron is employed, lest the fusibility of the cast iron should be diminished in consequence of the reduction in the proportion of carbon. It is necessary to employ coke as free from sulphur as possible, and a small quantity of lime is not unfrequently introduced with the iron, in order to combine with the silica contained in the coke, and prevent excessive waste of iron in the slag; but the lime is liable to corrode the lining of the furnace.

When casting in sand, it is necessary to prevent the direct contact of the melted metal with the sand, lest it should be chilled; for this purpose the *iron-moulder's blacking* is used, which is generally charred oak-wood ground to powder; the gas evolved by this under the action of the heated iron suffices to prevent the contact of the metal with the mould.

For some purposes it is found advantageous to melt the foundry iron with about one-third of scrap wrought iron.

In order to protect castings from rust, they are brushed over with linseed oil, and suspended over a very smoky wood fire, being afterwards dipped into turpentine, when they acquire a bright carbonaceous protective coating.

The following table shows the proportions of materials employed, to yield *one ton of cast iron*, in some of the most important British iron districts:—

	Staffordshire	Yorkshire	Scotland	South Wales
Ore .	40 to 60 cwt.	70 cwt.	36 cwt. (calcined)	67 cwt.
Limestone	15 to 18 ,,	20 ,,	10 ,,	17 ,,
Coal .	60 ,,	80 ,,	30 ,,	35 to 37 ,,

The iron made in Staffordshire is chiefly forge iron for conversion into common bar iron; Yorkshire produces forge iron for making best bar and plate; Scotland furnishes foundry iron; and Wales forge iron for conversion into railway bar iron.

CONVERSION OF CAST IRON INTO WROUGHT OR BAR IRON.

To obtain iron in a state fit for rolling into strong bars, boiler plates, &c., it is necessary to deprive the cast iron as far as possible of all foreign matters, except a small proportion of carbon, of which a quantity not exceeding $\frac{1}{200}$th is found to increase the toughness of the iron.

The chemical agent employed to effect the purification of the iron is oxygen, which is supplied by atmospheric air.

When iron is heated to redness in air, it combines with oxygen and becomes coated with *black oxide of iron*, which is detached in scales when the iron is struck with a hammer, and is more easily melted than iron. If iron containing carbon be strongly heated in contact with the black oxide of iron, the latter parts with its oxygen to the carbon, which it converts into carbonic oxide gas. If silicon be present in the iron, it combines with the oxygen to form silica; this unites with another portion of oxide of iron, yielding a *silicate of iron* which becomes liquid at a high temperature, forming a slag easily separated from the purified iron. Manganese combines with oxygen even more readily than iron, forming an oxide of manganese which is easily dissolved in the liquid slag.

The most important of the processes employed for the conversion of cast iron is that of *puddling*, but this was formerly often preceded by a *refining* process which will therefore be first described, although it is now comparatively seldom employed.

The *finery furnace* or *running out fire* (Fig. 10), in which the process of refining cast iron is carried on, is an oblong trough (A) made of cast iron, three sides of which are enclosed by double walls (U) which are kept cool by water circulating between them; it is about 3 feet long, 2 feet wide, and 2½ feet deep, the floor being rammed with sand and made somewhat concave. This trough is furnished with two, four or even six blast-pipes (*t*), similar in construction to those of the blast furnace, and inclined at an angle of twenty or thirty degrees to the floor of the hearth, so that the blast may be directed on to the metal. About 400 cubic feet of air per minute are usually supplied to the finery, giving a pressure of about 3 lbs. per inch. The hearth of the furnace having been filled with coke, four pigs of iron are arranged upon it, parallel to the four sides of the hearth, and two more are placed transversely over them; coke is heaped up over the pigs, and blown into a strong fire, when the cast iron gradually melts, and is partly converted into

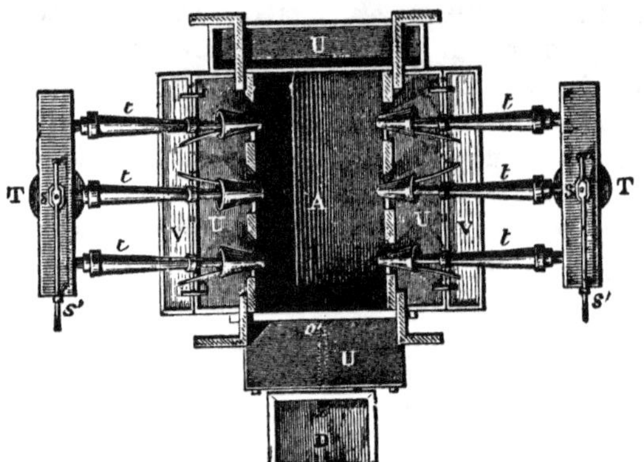

FIG. 10.—Hearth for refining Cast Iron, in section and plan. A, Hearth in which the fuel and pig-iron are placed. B, Hollow iron supports for the chimney C, which is about 18 feet high. D, Flat mould for the refined iron. *e*, Funnel-pipes for conveying water to the tuyères. *i*, Pipes for carrying off the heated water from the tuyères. *l*, Pipes for carrying off the heated water from the hollow walls U. *o*, Tap-hole for running out the fine metal. *r*, Stopcocks for supplying water to the cisterns *v*. *s*, Valves for regulating the blast issuing from the pipes T.

oxide by the blast of air; the oxide of iron imparts oxygen to the silicon present in the cast iron, and converts it into silica which combines with more oxide of iron to form the slag or *finery cinder*. The coke is replaced as it burns away, so that four or five hundredweight are used for refining a ton of pig iron. In about two hours, the tap-hole (*o*) is opened, and the *fine metal* or *plate metal* allowed to run into a shallow cast-iron mould (D), lined with loam and kept cool by the circulation of water beneath it. The metal is thus cast into a plate about ten feet long, three feet wide and two inches thick, whilst the bulk of the slag runs off into a separate mould beyond, some being left upon the surface of the fine metal. The latter is chilled by throwing water upon it, when it becomes very brittle, and is easily broken into masses fit for the puddling furnace.

The fine metal, as might be expected after the chilling, has much the character of white cast iron, but it has been purified from the greater part of its silicon, and the proportions of carbon, manganese, sulphur and phosphorus have been diminished. The phosphorus combines with oxygen to form phosphoric acid which is found in the slag as phosphate of iron. The sulphur passes into the slag as sulphuret of iron. When the refined iron is intended for the manufacture of the sheet iron with which tin plate is made, charcoal is employed in the refinery instead of coke which imparts sulphur to the metal, but the charcoal fire does not run the metal, which accumulates in soft masses; these are taken out and hammered into flat plates. 100 parts of pig iron yield only from 85 to 90 parts of refined iron, the loss representing the impurities which have been removed, and that portion of the iron which has been carried into the slag in the form of oxide.

One finery hearth is capable of refining ten tons of cast iron in twenty-four hours, so that it is able to keep pace with the blast-furnace.

In some works the pig iron is run into the finery hearth direct from the blast-furnace.

The slag from the refining process, or *finery cinder*, consists essentially of *silicate of iron*, composed, in 100 parts, of about 69 parts of oxide of iron (protoxide of iron or ferrous oxide) and 31 parts of silica, but it always contains sulphuret of iron and phosphate of iron, as well as manganese, magnesium and other metals in small quantity, which were present in the cast iron. As the 31 parts of silica contain 14½ parts of silicon, and the 69 parts of oxide of iron contain 53½ parts of metallic iron, it is evident that the silicon contained in the pig iron involves a loss of nearly four times its weight of iron in the slag.

The Puddling Process.—This process depends upon the same chemical principles as the refining just described, but differs widely from it in the mode of manipulation, the object of the puddler being to stir up together the melted or softened iron with melted oxide of iron, so as to ensure the contact of the latter with every portion of the metal upon which its purifying influence is to be exerted.

The furnace in which the puddling is executed is of the kind known as *reverberatory*,* so named from its having an arch which beats back the flame on to the iron to be heated.

This furnace is represented in Fig. 11, being built of fire-brick, and either tightly bound with iron bars, or even entirely cased with cast-iron plates, since it is required to maintain a very high temperature in the furnace, which would crack the brick-work unless it was well secured.

The grate F of this furnace is of unusually large dimensions, measuring about 4 feet by 3, so that the coal, which is the fuel employed, may give a large volume of flame to be drawn through the furnace by the draught of the chimney (C) and forced by the arch to play upon the metal on the hearth (A) before it escapes through the flue.

The hearth (A) is generally about 6 feet in length and 4 feet across the widest part opposite to the working-door (D.) Its

* From the Latin *reverbero*, to beat back.

construction varies much in different places, but it is generally made of cast-iron plates, supported upon cast-iron pillars, so that they may be kept below the melting point by the exposure of their under surfaces to the air. In order to prevent these plates from being rapidly corroded, they are commonly protected by a coating of slag. The *fettling*, or

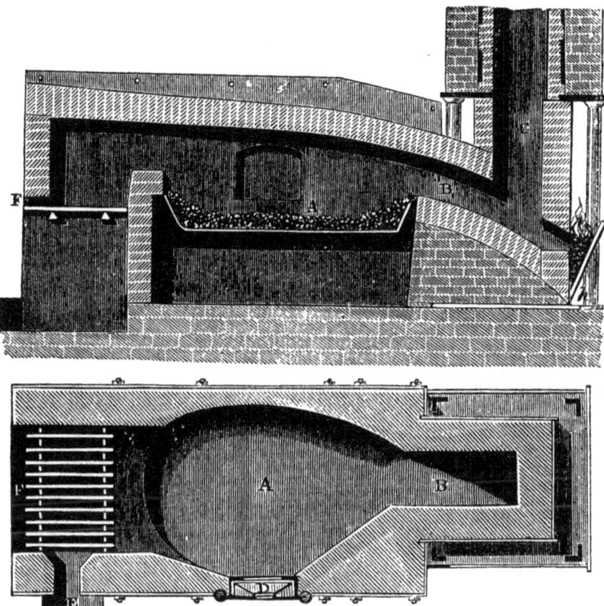

FIG. 11.—Puddling Furnace.

material with which the hearth is lined, has considerable influence on the puddling process. In some furnaces, *puddler's ore*, a variety of hæmatite, is employed, mixed with lime. Slag or *cinder* from the *re-heating furnace* is very commonly used. At the end of the hearth nearest to the grate, is a fire-bridge of brick, which prevents the coal from coming into contact with the metal on the hearth, and at the other end, near the chimney, a brick ledge, or *altar*, about 2½ inches high, prevents the iron from running out of the hearth. At this part, the width of the hearth is 2 feet. The height from the hearth to the highest part of the arch is 2 feet, but towards the chimney it is only 8 inches.

Beyond the altar is an inclined plane of fire-brick (B), down which the slag runs, being discharged from the furnace through the *floss-hole* (*h*), near which a small fire is maintained outside the furnace, to prevent the slag from solidifying.

The chimney varies from 30 to 50 feet in height, and is provided with an iron plate attached to a lever by which it may be raised or lowered at the will of the puddler, so as to increase or diminish the draught through the furnace.

The main working-door of the furnace, situated at the widest part of the hearth, is closed by an iron door suspended from a lever with a counterweight, so that it may be readily raised or lowered. In this door there is an opening about 5 inches square, through which the puddler thrusts the rake, *paddle* or *rabble*, with which the iron is stirred. Immediately beneath this door is another floss-hole stopped with loam, which can be opened at pleasure for the discharge of slag. Larger furnaces with two doors, enabling two puddlers to work at opposite sides, are now often employed.

The charging-door through which the metal to be puddled is introduced is situated at the cooler part of the hearth, near the chimney. The charge of a puddling-furnace consists of only 4 or 5 cwt. of pig iron or refined iron (usually mixed with grey pig iron), which is broken into fragments, and piled in heaps around the sides of the furnace. It is usual, especially when unrefined pig iron is being puddled, to add a quantity of hammer-slag or iron scales (black oxide of iron) in the proportion of about one-fifth of the weight of the metal, so as to obviate the necessity for allowing so large a quantity of the latter to combine with the oxygen of the air in order to form sufficient oxide to effect the purification of the metal. Even red hæmatite is sometimes added with the same object.

The charging door is now closed with an iron plate, and the grate is filled with coal, which is heaped up so as to close the fire-door completely. The damper at the top of the chimney being fully opened, so as to create a strong draught, the metal upon the hearth is soon brought to a high temperature, and in about twenty minutes it begins to melt. If it were allowed to become very rapidly liquid, the iron would be oxidised only to a slight extent upon the surface, and the object of the puddling process would be defeated; accordingly, a workman rakes the melting fragments into a cooler part of the hearth, and exposes fresh surfaces of the metal to the oxygen of the air in the furnace. In about ten minutes more, the whole of the metal has become fused to a pasty condition, and the skill of the puddler is now brought into play, to mix the semi-fluid metal with the melted oxide of iron upon its surface, by stirring it with a paddle introduced through the hole in the working door. At this stage much depends upon the consistence of the metal, which may be modified by judicious regulation of its temperature, for it is obvious that if the metal were in a perfectly thin and liquid condition, it could not be well incorporated with the oxide which is to purify it; the puddler therefore lowers the fire, partially closes the damper, and sometimes chills the metal by throwing water upon it. As the stirring or puddling is proceeded with, the metal froths or swells up very much, and evolves numerous bubbles of carbonic oxide gas, indicating the progress of the removal of the carbon from the iron. When unrefined pig iron is puddled, this *boiling* of the metal is more marked than in the case of refined iron, because the metal is more liquid, and hence the puddling of refined iron is sometimes distinguished as *dry puddling*, that of raw iron being called *pig boiling*. In a short time small clotted lumps of the purified iron separate or *come to nature* in the melted metal, the temperature of which is not high enough to fuse the iron when nearly deprived of its carbon, and at the end of an hour, so much purified iron has thus been separated that the charge works *dry* or *sandy*, and by this time the disengagement of carbonic oxide has almost ceased. It is now requisite to raise the temperature so as to soften the particles of purified iron, and allow of their being welded into a compact mass. With this view, the fire is made up, and the damper gradually raised until the particles of iron are so far softened that they stick together and cause the mass to *work heavy* under the paddle. A little of the soft metal is now collected upon the end of the paddle, and rolled about upon the hearth until a ball or *bloom* of about 60 lbs. weight has been collected; this is placed in the hottest part of the hearth, near the bridge, so as thoroughly to soften the metallic particles, which are then pressed together with the paddle in order to squeeze the slag from between them and to render the mass more compact. When the whole of the metal has been thus made up into (five or six) balls, the opening in the working door is stopped with coal, to prevent the entrance of any oxygen from the air, and the temperature is raised to a full welding heat. Each ball is then lifted out of the furnace on the end of an iron rod or *porter* which is pressed into it, and placed under a hammer capable of delivering about a hundred powerful blows in a minute, the ball being turned about under it, by the iron rod which serves as a handle, when the melted slag is forced out, in white hot showers, from between the particles of iron, and these become welded together into a *stamping*, or compact mass of metal of an oblong form which is rolled out into bars between the *puddling rolls*. These consist of a pair of massive iron cylinders (A A′, Fig. 12), the surfaces of which are so grooved that when the cylinders are placed together they exhibit a series of gradually diminishing openings, the first few being nearly oval, and the remainder (B B′) capable of rolling flat bars. The first two or three grooves are notched like files so that they may readily take hold of the metal presented to them, which is passed backwards and forwards by workmen stationed on opposite sides of the rolls, being sent through the first groove five or six times. These rolls, being made to revolve in opposite directions, powerfully compress the metal, and squeeze out any remaining slag. The bar is so turned about before being passed through each of the elliptical openings, that the pressure may be equally applied to all sides of it. So dexterous are the workmen employed in the rolling, that a minute and a half will often suffice for the conversion of the rough metal into a flattened bar, which is generally 4 inches wide, $\frac{3}{4}$ to 1 inch thick, and 10 or 12 feet long.

The rolls are generally cast in chill (p. 39), but so that the chill may not penetrate too deep and render the rolls brittle.

For producing bars of any special pattern, such as railway bars, rollers are employed which have grooves capable of imparting the required form to the bar passed between them. During the rolling of the bars, they are main-

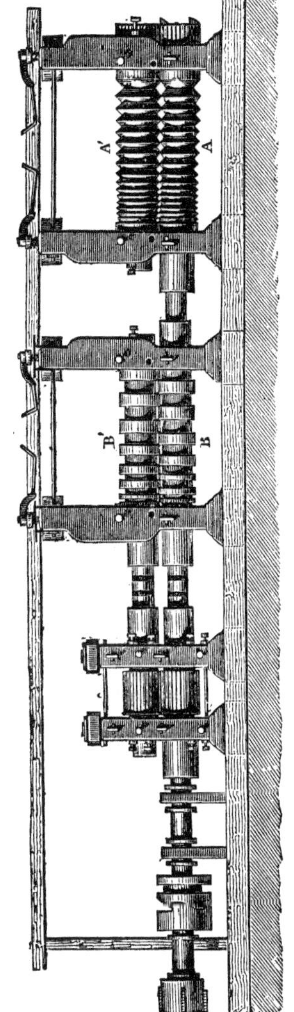

FIG. 12.—Mill for rolling Puddled Bars.

tained at a sufficiently high temperature by the heat which is extricated in consequence of the compression of the particles of metal. It is often necessary to allow a small stream of water to trickle over the rolls in order to prevent them from becoming over-heated and sticking to the metal.

In some ironworks, the compression of the puddled balls is effected between the jaws of a powerful squeezer (Fig. 13) worked by steam power, but a ponderous Nasmyth's steam hammer is now commonly employed, its invention having very materially assisted to develop and improve the art of forging iron. In others the puddled balls are carried at once, by tongs, to the bloom-squeezers (Fig. 14), a succession

FIG. 13.—Squeezer for Puddled Balls.

of three pairs of heavy rollers, of which the upper pair are grooved longitudinally to bite the bloom. On issuing from these the compressed bloom is caught by a projection attached to an endless chain by which it is lifted up to the platform in front of the roughing-rolls, when it is seized by the workman and forced into the largest groove.

Another machine for compressing the blooms consists of a massive iron table (B, Fig. 15) with a strong elliptical ledge (E), within which an iron cylinder (C) is made to revolve, so that a much wider interval (A) is left between its circumference and the ledge at one end than at the other; the bloom (D) being placed at the wider end, is carried round as the cylinder revolves, being well squeezed in its progress, and

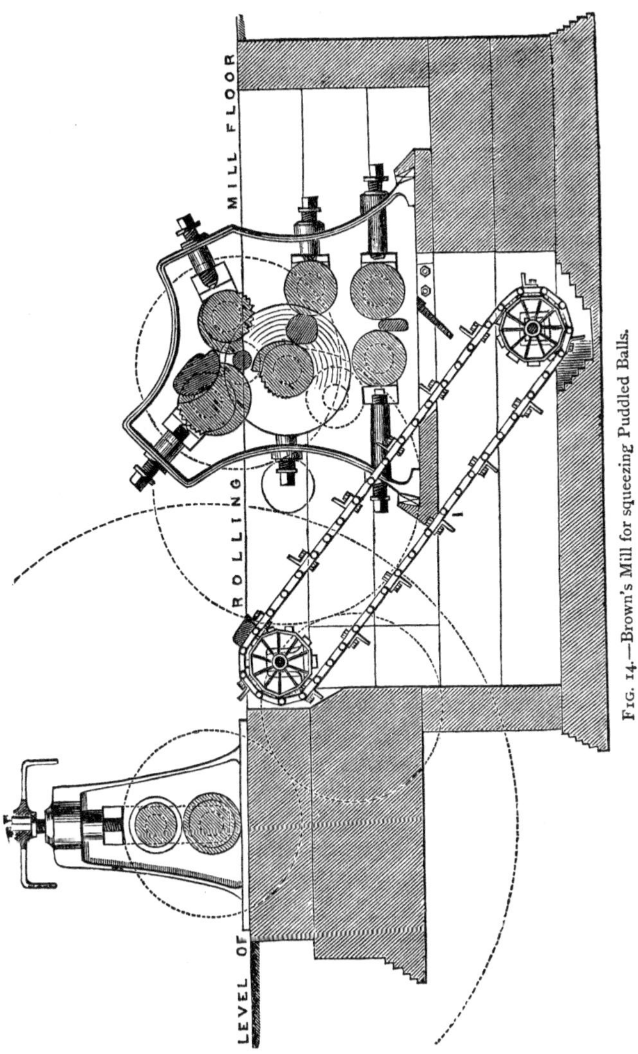

Fig. 14.—Brown's Mill for squeezing Puddled Balls.

delivered from the narrow end fit for passing through the rolls.

The process of puddling occupies about two hours, so that it is necessary to provide five puddling furnaces in order to work up the iron furnished by one blast-furnace and one refinery. The coal consumed is about equal in weight to the metal puddled.

From 90 to 92 parts of puddled bars are made from 100 parts of fine metal by a skilled puddler, but when the process is badly managed, a larger proportion of the iron is oxidised and carried off in the slag.

The slag or *tap-cinder* from the puddling furnace is often termed by chemists a *highly basic silicate of iron*, for it contains a very large proportion of oxide of iron and a small proportion of silica, but it always contains much phosphorus and sulphur extracted from the iron, these being present, respectively, in the forms of phosphate and sulphuret of iron. The results of the analysis of a sample are subjoined.

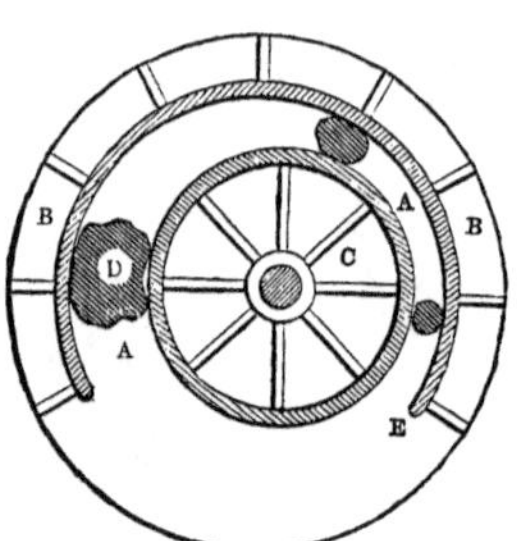

FIG. 15.—Cylindrical Rotating Squeezer for Puddled Balls, seen from above.

Composition of Tap-Cinder from Puddling Furnace.

{ Iron		54·33
{ Oxygen		16·87
Silica		8·32
Phosphoric acid	{ Phosphorus	3·18
	{ Oxygen	4·11
Sulphuret of Iron	{ Sulphur	2·57
	{ Iron	4·50
Lime		4·70
Oxide of Manganese		0·78
Magnesia		0·26
		99·62

The lime in this sample is due to the employment of that

substance as a lining for the hearth of the puddling furnace in order to assist in the removal of the sulphur.

It is evident that the waste of iron in the process of puddling will be greater in proportion as the crude metal is richer in silicon, since this element requires at least four times its weight of iron to be added in the form of oxide, in order to convert it into a fusible slag.

It has been observed that when the slag contains much manganese, the removal of the phosphorus and sulphur from bar iron is more complete, though the puddling is then attended with more difficulty. The phosphorus appears then to be converted into phosphate of manganese, from which the phosphorus cannot be again reduced and restored to the iron, as is the case when it is converted into phosphate of iron. The removal of the sulphur seems to be facilitated by the circumstance that the silicate of manganese formed in the slag has the property of dissolving the sulphuret of iron at a high temperature.

The extent to which the most important foreign substances are removed from the pig iron by puddling, may be inferred from the following comparison of the composition of a puddled bar with that of the good No. 3 grey cold-blast Staffordshire pig from which it was obtained:—

	Pig Iron	Puddled Bar
Carbon	2·28	0·30
Silicon	2·72	0·12
Phosphorus	0·65	0·14
Sulphur	0·30	0·13
Iron	94·05	99·31

Although, as a chemical operation, the puddling process may be regarded as effecting a very difficult purification by very simple means, as a manufacturing process it does not appear in so favourable a light. It demands from the puddler not only skilled labour, but labour of the most exhausting description, which must be rewarded by wages proportionally high. Exposed half naked to the heat of the furnace, and nearly blinded by its glare, the endurance of the puddler is tried to the utmost, and attacked very frequently by cataract and by pulmonary disease, he seldom lives beyond fifty, an age at which men labouring under happier conditions scarcely begin to show signs of decay. Nor does the expense of the process consist only in high wages, for the wear and tear of the furnaces, caused by the frequent repetition of such heavy work, is attended with a great outlay. The character of the malleable iron obtained, in so far as it depends upon the proportion of carbon remaining, is also very variable, from the want of any certain criterion by which the puddler may know when the process should be discontinued. A great many processes have been devised, from time to time, in order to dispense with the puddling operation in its present form. Revolving furnaces and other mechanical arrangements have been proposed, to replace the manual labour. The most recent and apparently the most successful of these is Dankes' furnace, which is a revolving cylindrical chamber lined with a mixture of lime with iron-ore free from silica, and supplied with air by a fan. The pig-iron is run, in a molten state, into the furnace, in charges of about 600 lbs. A quantity of oxide of iron is added (as in the puddling process), and the cylinder is made to revolve until the metal has come to nature, when it is removed in a single ball. The process is assisted by directing a jet of water against the lining on the descending side of the furnace, to chill part of the cinder and carry it under the iron.

The only process which has been attended by a sufficient measure of success to attract the general attention of ironmasters is that of Bessemer, which consists in blowing air through the melted metal in a huge crucible, so that the mechanical agitation caused by the blast may exert the same beneficial influence as the labour of the puddler. Since, however, this treatment does not effect the removal of the sulphur and phosphorus from the metal, it is not able to produce good bar iron from any but the best qualities of pig, such as are considered too expensive for conversion into bar. Its application in this country is therefore in great measure limited to the manufacture of Bessemer steel, and a description of the process will be given under the head of Steel.

In Sweden and India, where charcoal pig iron may be obtained at a cheaper rate, the process of Bessemer is extensively adopted, and although it is not carried out in precisely the form of apparatus described under *Steel*, there is no difference in the principle.

In Silesia the conversion of the cast iron into malleable iron is effected in a reverberatory furnace heated by the flame of a mixture of coal gas and air which are supplied to the furnace through two sets of tuyères. When the cast iron is melted on the hearth of the furnace, a little limestone is thrown in, and two strong blasts of air are directed upon the surface of the metal from separate tuyères, introduced through openings on each side of the hearth. The blast keeps the metal in constant motion, and blows away the slag from the surface so as continually to expose the metal, which is also occasionally stirred with a rake, a little limestone being occasionally added. Two tons of pig iron are treated in this furnace, and require from 2½ to 5 hours. The temperature in this furnace being much higher than that of the English puddling-furnace, the iron is run out in a perfectly liquid condition.

The ingots or bars cast from the melted malleable iron, whether obtained by this process or by that of Bessemer, require to be hammered and rolled in order to give them the usual strength of bar iron. Thus, a given specimen which was broken by a tension of 18½ tons upon the square inch when first cast into an ingot, required a tension of 32½ tons after being hammered and rolled.

Conversion of Puddled Bar Iron into Merchant and Best Bar.—The *Mill Bar* or Puddled Bar Iron obtained, as described above, by rolling the puddled balls, is of very inferior quality. Its tenacity is so low that it is quite unfit for axles and portions of machinery which have to resist any strain, and it could not be drawn into wire. Some specimens of puddled bar have a tensile strength as low as 9 tons per square inch, whilst the best varieties of bar iron have a strength of upwards of 25 tons. The fracture of puddled bar exhibits a coarse crystalline structure, which is generally found to indicate inferior tenacity. The hardness of puddled bars, however, is greater than that of the superior varieties of bar iron, and this quality recommends it for such applications as the construction of railway lines where high tenacity is not of so great importance. In order to obtain a bar iron of better quality, the puddled bars are cut up, while hot, by large shears, into lengths of a foot or more, according to the length of the bar required, and are submitted to the process of *fagotting* or *piling*. This consists in placing four of the lengths upon each other, so as to form a pile; this is placed in a *re-heating* or *mill-furnace*, raised to a welding heat, and passed between rollers which convert the four bars into a single bar, to be drawn out, as before, to the required dimensions. The *merchant bar* thus obtained may be still further improved in quality by a repetition of the process of rolling, which is generally executed upon the bars doubled upon themselves, producing the toughest description of wrought iron, known as *best bar* or *wire iron*. Such iron may be bent double, in a cold state, or a bar of it may be tied into a knot when cold without exhibiting a crack. In piling iron for boiler plates, cannon, &c., the bars are sometimes placed crosswise over each other in the pile, so that the fibres of one layer may be at right angles with the fibres of the next layer.

The *Mill-furnace* or *Reheating-furnace* (Fig. 16) employed for raising the bars to a welding heat, is a reverberatory furnace so constructed that as little air as possible shall reach the hearth without having previously been deprived of oxygen by passing through the fuel, for otherwise a large proportion of the iron would be wasted in the form of oxide. The furnace has only one opening to the hearth, through which the piles of iron to be fagotted are introduced; this opening is situated immediately under the chimney, so that when the door is opened in order to introduce the iron, the air which is drawn through it passes up the chimney instead of sweeping over the hearth of the furnace.

Experiments upon a particular sample of puddled bars, having a tensile strength of 19½ tons per square inch, showed that its strength increased progressively during five processes of fagotting, till it eventually reached 27½ tons, when a repetition of the process was found to weaken it again,

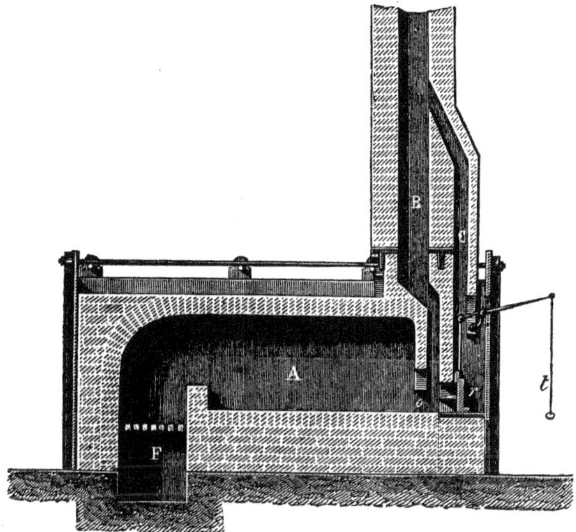

Fig. 16.—Reheating Furnace. A, Hearth upon which the bars are heated. F, Grate. r, Door opened by the lever t, for introducing the bars. o, Opening through which the flame and smoke pass into the flue B. C, Flue through which the air entering by the charging door passes into the chimney.

until, after the eleventh fagotting, it was found to have returned to its original strength of 19½ tons.

The coal consumed in producing one ton of merchant bar iron from the ore, amounts in all to about five tons, and every reheating for the process of piling involves the consumption of a quantity of coal equal to about half the weight of the iron.

The great improvement in the strength of malleable

iron by the processes of fagotting and rolling has been more satisfactorily established by experience than explained by theory. One obvious effect of the violent compression between the rollers is the squeezing out of slag, which is liable to become entangled in the iron during the hammering and rolling of the balls taken from the puddling-furnace. The occurrence of small masses of slag in malleable iron is not an uncommon cause of weakness, each particle of slag giving rise to a flaw in the metal. In the process of reheating the bars this slag is melted, and may then be squeezed out by the action of the rollers.

A marked diminution in the proportions of carbon and silicon present in the iron is also effected during the process, as shown by the following results of chemical analysis :—

In 100 parts	Carbon	Silicon
Puddled Bar . . .	0·296	0·120
Best Bar	0·111	0·088

This may be explained by the action of the oxide of iron formed upon the surface of the bar during exposure to air at a welding heat (see p. 49).

The rolling of several bars into a single bar would render the structure of the metal uniform, so that the bar would be equally strong throughout.

During the operations of fagotting and rolling, the iron acquires a remarkable fibrous structure, so that if a bar of the best iron be notched with a chisel, and broken across by a steady pressure, the fracture will present a stringy appearance, resembling that of a green stick; whilst a puddled bar thus treated would exhibit a crystalline shining fracture, not unlike that of cast iron. That this *nerve* or *reed*, as the fibrous structure is sometimes called, should materially increase the resistance of a bar to any transverse strain, can readily be believed, for such a bar resembles a bundle of wires firmly bound together, whilst a crystalline bar must be regarded as composed of a number of particles of iron stuck together in a confused manner. But with our present imperfect acquaintance with the mutual relations and movements of the individual particles composing a solid mass, it is not easy to give a satisfactory explanation of the production of the fibrous structure by rolling the softened bars in the direction of their length. Much less can we explain the circumstance, which appears to have been satisfactorily established, that this fibrous structure is liable to re-conversion into the crystalline structure if the iron be subjected to a long succession of powerful vibrations.

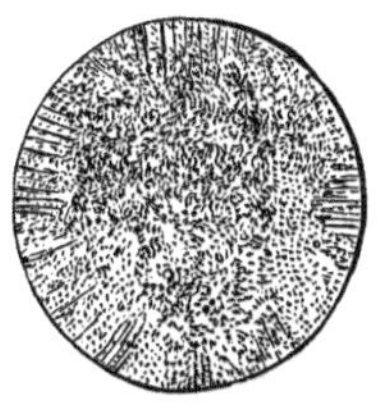

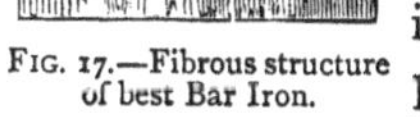

Fig. 17.—Fibrous structure of best Bar Iron.

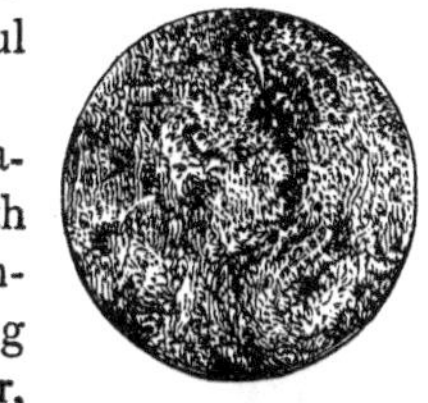

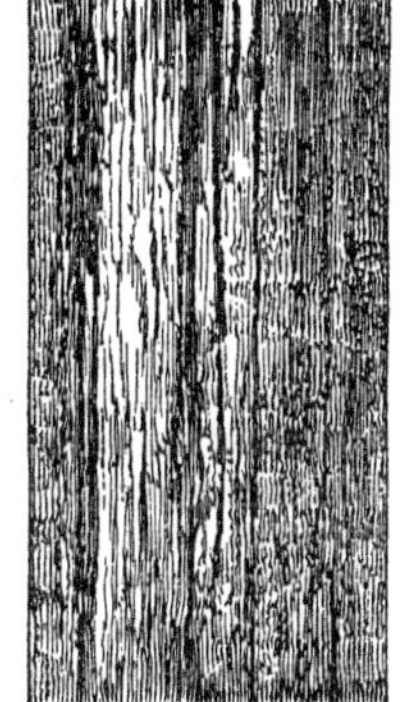

Fig. 18.—Structure of inferior Bar Iron.

The deterioration in the strength of bar iron by often-repeated forging under the hammer, is commonly explained as resulting from this change in structure; and axles, girders, &c., originally made of fibrous iron, are said to have snapped unexpectedly, exhibiting a crystalline structure. Hence, in cases where the iron is to be exposed to much vibration, a fine-grained wrought-iron richer in carbon is preferred to a fibrous iron.

In drawing any inference as to the quality of wrought iron from the character of its fracture, it is most important that the mode of breaking it should be taken into account, for it is found that a bar or plate which exhibits a fine fibrous structure when broken by bending, appears crystalline when suddenly snapped, or when broken by a blow from a shot; and it is probable that a want of attention to this has given rise to many of the contradictory statements with respect to alterations in the structure of wrought iron under various conditions.

Some of the recent experiments upon this subject lead to the conclusion that the fibrous fracture of the best descriptions of bar iron is due to their ductility, which causes them to draw out before breaking, whereas an inferior bar snaps without drawing out, the suddenness of the fracture preventing the appearance of fibre.

When a bar of the best iron is soaked for a few days in

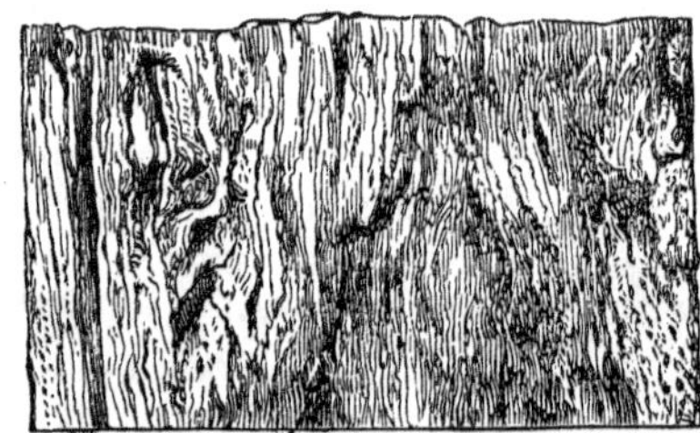

FIG. 19.—Structure of Puddled Bar.

very weak muriatic or nitric acid, its fibrous structure is rendered manifest, and it presents the appearance of a bundle of fine straight fibres laid regularly side by side (Fig. 17). In an inferior bar (Fig. 18) the fibres are coarser and not so regularly arranged; whilst in a puddled bar (Fig. 19) there is little or no regularity of structure disclosed.*

* The circumstance that the fibre resists the action of the acid longer than the metal lying between the fibres, would lead a chemist to suspect that the fibres consist of crystallised, and the interstitial portion of amorphous (or *colloid*) iron. It is a common observation in the laboratory that when one portion only of a mass is crystallised, it is much less readily melted or dissolved than the portion which is not in that condition. It is also frequently seen that the uncrystallised portion of such a mass is converted into a mass of crystals, in course of time, or if briskly shaken or stirred. How far is the alleged change in the structure of bar iron from fibrous to crystalline really caused by the crystal-

Bar iron which cracks when doubled or tied in a cold state is said to be *cold-short*; and this defect is very frequently due to the presence of phosphorus in the metal. Bar iron commonly contains about one-thousandth of its weight of phosphorus, without injury to its strength, and, it is said, with an improvement in its capacity for welding; but when the proportion of phosphorus amounts to one part in two hundred of iron, its effect in diminishing the tenacity of the iron becomes perceptible; and when it reaches one and a half in two hundred, it renders the iron decidedly cold-short.

The cold-short iron loses its brittleness at a high temperature, so that it may be forged with great readiness.

The fracture of iron containing so large an amount of phosphorus is commonly found to present a coarsely crystalline structure.

Silicon is said to injure the tenacity of bar iron, whether hot or cold, in a higher degree than phosphorus; and arsenic, which is occasionally present, renders the iron cold-short and hinders welding.

When the iron cracks while being worked under the hammer at a welding heat, it is termed *hot-short* or *red-short*, and this quality is often exhibited by iron which is sufficiently pure to be very tough when cold. The presence of sulphur has a decided effect in causing red-shortness, even if there be only four parts in ten thousand parts of iron; and it is alleged, with much probability, that enough sulphur for this purpose is sometimes imparted to the iron by the products of combustion of the coal in the reheating furnace. Red-shortness appears to be sometimes caused by a deficiency of carbon in the bar-iron. *Burnt iron*, which is so red-short that it cannot be welded, is believed by some metallurgists to have lost too much carbon in the processes of puddling and forging, while others attribute its defects to the presence

lisation of the *colloid* portion of the metal, which would have the effect of rendering it less plastic or ductile, and more liable to *snap*, a mode of fracture which would prevent the fibrous structure from becoming visible?

of oxide of iron formed by the action of the air upon the heated metal.

There appears to be little knowledge of a thoroughly satisfactory character with respect to the effect of different proportions of foreign matter upon the quality of malleable iron, for the exact analysis of this material is tedious and difficult, and those who are competent to execute it in a trustworthy manner have rarely the opportunity of becoming practically acquainted with the behaviour of the metal in the workshop and the forge.

STEEL.

The difference between steel and wrought iron (or soft iron, as it is sometimes called) is chiefly seen in their behaviour when raised to a high temperature and suddenly cooled by being plunged into water; when wrought iron undergoes little if any change; while steel is rendered almost as hard as diamond, and so brittle that it snaps off if an attempt be made to bend it. If this very hard brittle steel be again heated to a temperature far short of redness, and cooled, it becomes much softer than before and extremely elastic, so that when forcibly bent it springs back into its former position, whereas wrought iron would retain a permanent bend. It will be remembered that cast iron is also greatly hardened by being suddenly cooled, but it cannot be rendered elastic like steel.

The property of being hardened by chilling is dependent upon the presence of carbon in the metal, for chemically pure iron does not exhibit it, though all specimens of commercial wrought iron are slightly hardened when so treated, because they contain a small proportion of carbon. It is, therefore, difficult to define the exact limit beyond which wrought iron passes into steel. Bar iron, containing as much as two parts of carbon in a thousand of metal, would be so decidedly hardened by chilling as to be termed a *steely iron*, and a slight increase in the quantity would produce a *mild steel*, such as the *homogeneous metal* of which cannon and armour-plates are forged. A proportion of carbon, amounting to three parts in the thousand, is contained in the Bessemer steel rails, and in the steel of which spades and hammers are commonly made. Twice this proportion of carbon, or six parts in a thousand, is contained in steel ramrods; and the steel employed for tools commonly contains ten or twelve parts of carbon in a thousand. When the carbon amounts to fourteen parts in a thousand, the steel becomes more fusible and resembles white cast iron.

Bar iron, which contains only a minute proportion of carbon, has a tensile strength of about 57,500 lbs. (nearly 26 tons) per square inch*; but when the proportion of carbon amounts to three or five parts in a thousand, its tensile strength is increased to 90,000 or 100,000 lbs. (40 tons or 45 tons) per square inch, while it is still soft enough to be easily punched and flanged. Such a metal is well suited for boiler-plates and similar purposes.

For armour-plates designed to protect ships, by offering resistance to blows from shot at high velocities, it appears to be desirable that the proportion of carbon should not exceed two parts in a thousand, for, although an increased proportion of this element is attended with increased resistance to tension (or to a force tending to pull the particles of the metal asunder), as well as to penetration by dead pressure, it renders an armour-plate more liable to fracture under the sudden and powerful blow of a shot.

In addition to hardness and increased tensile strength, the presence of carbon confers another valuable property upon the iron, namely, the capability of retaining magnetism. When a bar of soft iron is placed in contact with a magnet, or in the axis of a coil of wire through which a galvanic current is transmitted, the iron bar acquires all the properties of a magnet, becoming capable of attracting iron filings, and, if freely suspended, of pointing, like a compass needle, north and south. It loses these magnetic properties as soon

* The Board of Trade gives 5 tons per square inch as the *working strain* tc which it is safe to expose bar iron in actual practice.

as the permanent magnet is removed, or the galvanic current is discontinued in the surrounding wire. A bar of steel, however, would retain its magnetic properties. Just as all specimens of commercial bar iron are slightly hardened by chilling, in consequence of their containing a minute quantity of carbon, so they are all slightly retentive of magnetism in proportion to the quantity of carbon present. In the manufacture of electro-magnetic instruments of various kinds, the action of which depends upon the sudden loss of magnetic power by a bar of iron when the galvanic current is interrupted, it becomes of great importance to select the

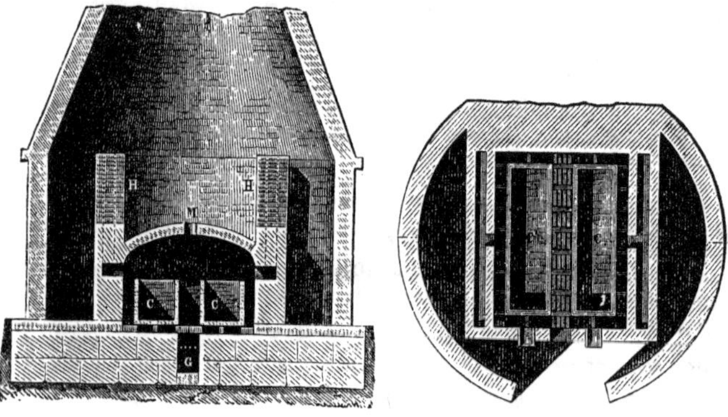

FIG. 20.—Cementation Furnace for converting Bar Iron into Steel.

softest commercial iron, so that it may lose its magnetic power as quickly as possible. On the other hand, the hardest steel is selected for the production of permanent magnets, and it has been noticed that the addition of the metal tungsten to steel increases its power of retaining magnetism.

Steel rings when struck, much more than iron, and this property is relied upon by the steel-maker as one of the tests of its quality.

Conversion of Bar Iron into Steel by Cementation.—The process of *cementation* by which until lately nearly all English steel was produced, consists in heating bar iron in contact with charcoal, in a closed chest, until it has acquired a proper proportion of carbon.

The cementation furnace (Fig. 20) is dome-shaped, like the furnace of a glass-house, and is enclosed in a conical jacket of brick-work which serves to carry off the smoke from the flues. The hearth of the furnace is divided into two parts by the grate G, traversing the whole length (13 or 14 feet) of the furnace, in which a coal fire is maintained, the flame of which is made to circulate above, below, and around the fire-clay chests or *pots*, or *troughs* C placed one on each side of the grate, before escaping through the flues in the wall H and through the opening M. These troughs are 10 or 12 feet long, and about 3 feet in depth and width, so that each will contain seven or eight tons of bar iron, together with the charcoal necessary for its conversion into steel. A small opening is left at about the middle of one end of each chest, through which the end of one of the bars undergoing cementation is allowed to project; this *proof-bar* is withdrawn from time to time, through a small door in the wall of the furnace, for the purpose of watching the progress of the cementation. There is also a small door in the wall of the furnace, a little above the top of each trough, through which the bars of iron may be introduced and withdrawn, a larger door being made in the middle of the wall to allow the passage of the workmen.

The nature of the *cement* or carbonaceous material employed varies in different works, but the bulk of it nearly always consists of ground charcoal from hard wood, with which there are sometimes mixed a little common salt and some ashes of the charcoal. The salt and the alkaline matters contained in the ashes are believed by some persons to have a beneficial effect in converting into a glass the silica which is contained in the charcoal, and thus preventing it from imparting silicon to the steel. The bars of iron should be of the purest description if the best steel is to be produced. They are about three inches broad, and one-third of an inch thick.

In order to fill the troughs, the workman stands upon an iron platform between the two, and sifts the cement powder into them, so as to form a layer of about half an inch in depth, upon which the bars are arranged, standing upon their edges, at about an inch apart. More cement powder is now sifted over these, so as to fill up the intervals between them, and to cover them entirely to the depth of about an inch. Upon this a second layer of bars is placed, then more of the charcoal powder, and so on until the trough is filled to within a few inches; it is then covered in with four or five inches of fire-clay or some similar material, well rammed down, and the fire is gradually applied during the first two or three days, to avoid the risk of splitting the troughs. A temperature high enough to melt copper (estimated at about 2,000° of Fahrenheit's scale) is required to enable the bar iron to acquire a proper proportion of carbon, and the troughs are maintained at this temperature for a period proportionate to the hardness which the steel is required to possess; four days being sufficient for producing the steel of which saws and springs are made, while six or eight days are required for shear steel, and ten days or more are required for the very hard steel of which cold chisels are made. The fire is then gradually let down, to avoid sudden change of temperature, so that some days elapse before the troughs are cool enough to be opened. About three weeks are commonly occupied in the conversion of the bar iron into steel—one to get up the heat, one to keep it at the required degree, and one to cool it down; so that only about sixteen cementations can be executed in a year by a single furnace.

The bars are found to have upon their surface bubbles or blisters of considerable size, whence they are called *blister steel.* On breaking the bars, the fracture exhibits a silvery lustre and a well-marked crystalline structure. The proportion of carbon which has entered into combination with the iron depends upon the duration of the cementing process, but it rarely exceeds fourteen parts in a thousand parts of the metal.

The chemical changes which are involved in the process

of cementation are not yet thoroughly understood. The passage of the infusible solid carbon into the interior of the solid iron bar obviously requires explanation. It might be imagined that the external particles of iron which are in contact with the charcoal, becoming surcharged with carbon, impart a portion of that element to the next layer, and so on, until the particles in the very centre of the bar had acquired a proper share of carbon; but such an explanation would require that the outside of the bar, at the close of the process, should be very much richer in carbon than the inside, and we have no evidence that this is the case.

The following explanation appears more probable. The small quantity of oxygen contained in the air remaining in the trough, and present in the pores of the charcoal, enters into combination with the carbon to form carbonic oxide gas; this gas, in contact with iron at a high temperature, gives up one-half of its carbon to the metal, and becomes converted into carbonic acid gas; but this carbonic acid, in contact with the strongly-heated carbon, is reconverted into carbonic oxide, which again transfers one-half of its carbon to the metal, these changes recurring many times in the same order until the whole of the iron is converted into steel. The observations of chemists during the last few years have shown that red-hot iron allows the passage of gas through its substance, and that this metal has the power of absorbing a considerable quantity of carbonic oxide, which renders it easy to account for the transference of carbon from the charcoal into the interior of the bar.

Other gases containing carbon are capable of imparting that element to iron; thus, if coal gas, which contains carbon in combination with hydrogen, be passed for an hour through an iron tube containing some soft iron wires heated to bright redness, the wires will absorb carbon from the gas and become converted into steel.

The blisters, which are distributed sparsely and irregularly over the surface of the bars, are commonly believed to be due to the action of particles of oxide of iron, or of slag, accidentally occurring in the iron bars, upon the carbon combined with the iron, giving rise to carbonic-oxide gas, which escapes as a bubble through the softened iron.

As might be anticipated, the blistered steel, in its present condition, is only fitted for very rough articles, such as shovels; its largely crystalline structure renders it deficient in tenacity, and the bars are further weakened by their want of uniformity and by the presence of the blisters.

Conversion of Blistered Steel into Tilted or Shear Steel.—The quality of the blister steel is improved by a process similar in principle to the fagotting of bar iron. Five bars of blister steel are bound together into a bundle, being secured by a stout steel wire; four of the bars are about 18 inches long, and the fifth is twice that length, so that it projects beyond the others and forms a handle. This bundle is raised to a welding heat in a forge, sprinkled with sand to combine with the oxide of iron and form a fusible slag (see p. 49), and placed under the *tilt-hammer*; this hammer weighs about two hundredweight, and it is so suspended that it may be raised by *cams* projecting from the circumference of a wheel, the revolutions of which bring them down in succession upon the tail of the hammer, the head falling again upon the anvil as soon as the cam has passed. A few blows from this hammer soon weld the bars together, when the binding ring is knocked off, the bundle again heated in the forge, and hammered or *tilted* throughout its whole length, and on all sides, until it is reduced to a rectangular bar of the right dimensions. In order to avoid the necessity for reheating the bar during the process, the tilting must be effected with great celerity, and the hammer is made to deliver 300 or 400 blows per minute, the number being, of course, regulated by the rate at which the cam-wheel revolves. The workman being seated upon a swinging bench which brings him upon a level with the anvil, is enabled to move to and fro with little effort, and to bring every part of the elongated bar under the strokes of the hammer. The fracture of a bar of shear steel shows it to possess a much more compact structure than the blister steel, and its tenacity and ductility have been much improved by the tilting. It is probable also that, as in forging bar iron, the proportion of carbon has undergone a slight diminution, the steel being found to become softer after repeated tilting. If *double shear steel* be required, the tilted bar is broken and the two pieces welded into a single bar.

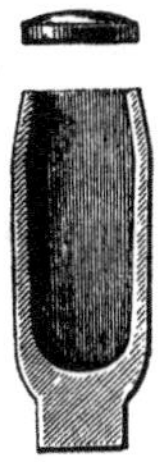

FIG. 21.

Shear steel derives its name from its use in making certain large shears, and it is commonly employed for tools which are to possess considerable toughness, without being extremely hard, such as scythes, plane-irons, and large knives; but its deficient hardness prevents it from taking a very high polish or a very keen edge, so that it will not serve for the finer kinds of cutlery. The best variety of steel used for these is made by melting the blister steel and casting it into ingots.

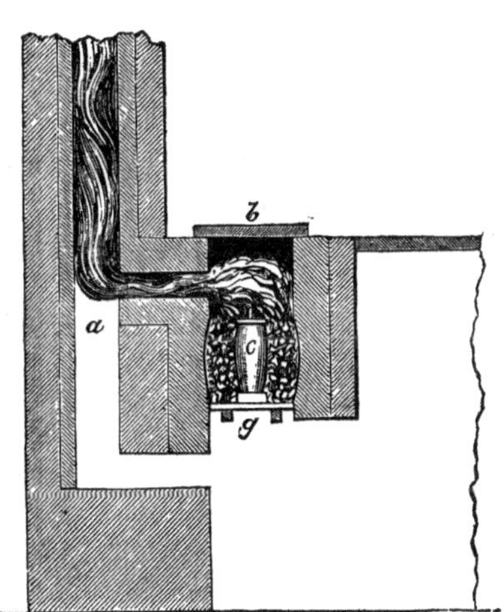

FIG. 22.—Furnace and Pot for melting Steel. *g*, Grate. *c*, Crucible. *b*, Cover of Furnace. *a*, Chimney.

Conversion of Blister Steel into Cast Steel.—The blister steel is broken up into pieces of a convenient size for packing close together, and about 30 lbs. of it are introduced into a tall narrow crucible (Fig. 21), about two feet high, made of fireclay mixed with black-lead, and provided with a closely-fitting cover. Some steel-makers add a little bottle-glass to fuse over the surface and prevent oxidation of the steel.

The crucibles are placed in a small furnace (Fig. 22) holding six, twelve, or more, about one foot wide and two feet deep, the opening of which is usually on a level with the floor, to facilitate the lifting of the crucibles. Several of these furnaces are connected by flues with the high chimney of the works, so that a powerful draught may be produced. Hard coke broken into small pieces is employed to raise the crucible to a bright red heat; the steel is then introduced, the crucible covered, and the furnace filled up with coke; when the steel is melted, the crucible is lifted out with a pair of tongs, and its contents poured into a rectangular or octagonal mould of cast iron which has been previously heated and is placed vertically for the steel to be poured in. The mould is made in two halves, closely fitting together, so that it may be opened for the removal of the bar of cast steel, and is coated inside with coal-tar soot. The quality of the cast steel produced is in some measure dependent upon the temperature at which it is poured, so that an experienced workman is employed for the purpose.

For the production of large castings of steel in this way, the requisite number of crucibles must be emptied into the mould as nearly as possible at the same time. At the factory of Krupp at Essen, near Cologne, a casting of 16 tons may be produced in this way, 400 men being well drilled to co-operate in emptying 1,200 crucibles, so that the melted steel may flow in an uninterrupted stream along the gutters leading to the mould. Great alterations and improvements in the manufacture of cast steel will probably result from the introduction of the regenerative gas-furnace of Siemens, in which steel may be easily melted in large quantities.

Cast steel is much more uniform in structure than tilted steel, and has a very compact granular texture, without lustre, indicating high tenacity, as may be seen on inspecting its fracture; though here, as in the case of bar iron (p. 67), it must be borne in mind, that when fracture takes place slowly, it will present a more or less distinctly fibrous appearance; the granular structure becoming evident only on sudden fracture. The higher the quality of the steel, the finer is the granular structure exhibited by sudden breaking. The lower qualities somewhat resemble bar iron in fracture. When produced by the process just described, cast steel has the serious defect of being brittle at a high temperature, so that it is forged with difficulty, and does not admit of being welded. But a method of correcting this was patented by Heath, in 1839, which consists simply in adding to the cast steel, in the melting-pot, about one-hundredth of its weight of *carburet of manganese*, the result of the action of heat upon a mixture of black oxide of manganese (ore of manganese) and charcoal, or some other substance containing carbon, such as coal tar. After this addition, the cast steel possesses much more tenacity at a high temperature, and can be welded either to itself or to wrought iron, so that it may be employed for the fabrication of many implements which were formerly obliged to be made of shear steel. Thus, the blades of table knives can be made of cast steel welded on to an iron *tang*, as that part of the knife is called which is fixed into the handle.

Another important consequence of the introduction of Heath's process has been a reduction of about one-third in the price of cast steel, by enabling it to be produced from an inferior quality of bar iron, instead of the very expensive descriptions which it was necessary to employ previously to this discovery.

The mode in which this addition of manganese acts to produce so great an improvement in the quality of the cast steel is by no means understood. It does not appear to depend upon the formation of an alloy of manganese with the steel, for the bulk of that metal is found in the slag from the melting-pots, only a minute proportion entering into the composition of the steel; but that even this small quantity affects the quality of the metal, appears to have been proved by the observation that the manganiferous steel, if melted a second time, becomes as red-short as if no manganese had been added, probably because the small proportion of that metal is removed by the oxygen of the air during the re-melting. It is commonly believed that manganese has a particular tendency to encourage the removal of sulphur, phosphorus, and silicon in the slag both from steel and iron, though its precise mode of action has not been defined (see p. 61).

For the manufacture of some tools requiring rough usage, such as the chisels of planes, it is customary to employ a bar of iron faced with steel, the cutting edge being, of course, made upon the steel side, which receives great support from the wrought iron. To produce such a compound bar, a bar of iron is polished upon that surface which is to be faced with steel, heated to redness, sprinkled with borax to cleanse the oxide from its surface, and placed in the ingot mould destined to receive the cast steel, which then adheres firmly to the polished iron, so that the two may be forged together.

Production of Bessemer Steel.—It has been already stated that Bessemer's process for converting cast iron into malleable iron depends upon the removal of the carbon by forcing air through the liquid metal, and if this process be arrested before the removal of the carbon is completed, the metal will have the composition of steel. Another process by which this kind of steel is produced consists in depriving the cast iron of nearly the whole of its carbon, so as to obtain wrought iron, which is then converted into steel by adding the proper proportion of carbon in the form of *Spiegel-eisen.*

The *converter* (Fig. 23) in which this process is carried out is made of wrought iron boiler-plate, and lined with fire-clay or other refractory material to protect it from oxidation. It is sometimes large enough to contain ten tons of cast iron for a charge, and is suspended on trunnions so that it may be easily tilted for charging and discharging. A six-ton converter is 11 feet high and 5½ feet in its widest diameter.

Through the bottom of this vessel there are several openings to admit the blast of air, which is blown in at a pressure of fifteen or twenty pounds upon the inch through 35 holes, from 7 tuyères with 5 holes each. (The number of tuyère holes employed varies, in different works, between 35 and 150, their diameter varying from $\frac{1}{4}$ to $\frac{3}{8}$ inch.) The converter having been heated by burning a little fuel within it, is charged with pig iron which has been previously melted in a separate furnace, a pig iron containing a large proportion of graphite and silicon and a small proportion of sulphur and phosphorus being selected. The air, bubbling through the liquid metal, induces an intense combustion of the iron, producing a large quantity of the black or magnetic oxide of iron which is carried up by the force of the blast together with the nitrogen of the air, which does not act upon the iron. The bubbles of this gas being forced up through the melted metal, effectually mix the unoxidised portion with the melted oxide, which converts the carbon of the cast iron into carbonic oxide gas, and the silicon into silicic acid, the latter combining with some oxide of iron to form a slag which appears as a froth at the mouth of the converter. The silicon is always oxidised before the carbon, and during the first ten minutes or so, very little flame appears at the mouth of the converter. Since, in this process, a portion of the iron itself is the fuel undergoing combustion, the temperature is much higher

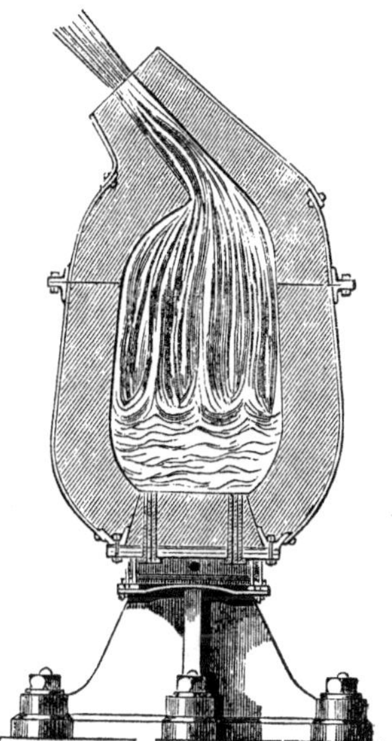

FIG. 23.—Bessemer's Converting Vessel.

than that of the puddling furnace, in which coal is the fuel, for a given quantity of oxygen, in the act of burning iron, produces above one-third more heat than in the act of burn. ing carbon. The temperature produced in the converter is able to effect the complete fusion of the purified iron, which remains liquid throughout the operation instead of separating in a pasty form as in the process of puddling. The manifestation of energy during the conversion is very striking; the roaring of the blast in passing through the molten iron, the long flame of the carbonic oxide, variegated by the combustion of small quantities of metals, the brilliant scintillations from the iron, and the white hot flakes of slag whirled upward by the blast combine to produce a volcanic effect which is not easily forgotten.

The operation usually lasts for only twenty minutes, its termination being indicated by the almost total disappearance of the flame of carbonic oxide, but a far more exact method of ascertaining when the requisite amount of carbon has been removed, consists in viewing the flame through an optical instrument known as the *spectroscope*, which enables the observer to detect certain lines in the spectrum or image of the flame, the disappearance of which lines marks, to within a few seconds, the conclusion of the process.

If Bessemer iron were required, the contents of the converter would now be discharged into a ladle (Fig. 24) and thence into moulds having the form of the required bars, but it has been already explained that the necessity for employing a high-priced pig iron prevents the economical application of this otherwise excellent process to the production of malleable iron in this country.

In order to convert the decarbonised metal into steel, the requisite proportion of carbon is added in the form of *Spiegeleisen* or *specular iron* (which must not be confused with the specular *ore*). This may be defined as a special variety of white cast iron containing a large quantity of carbon in chemical combination, together with much manganese. It

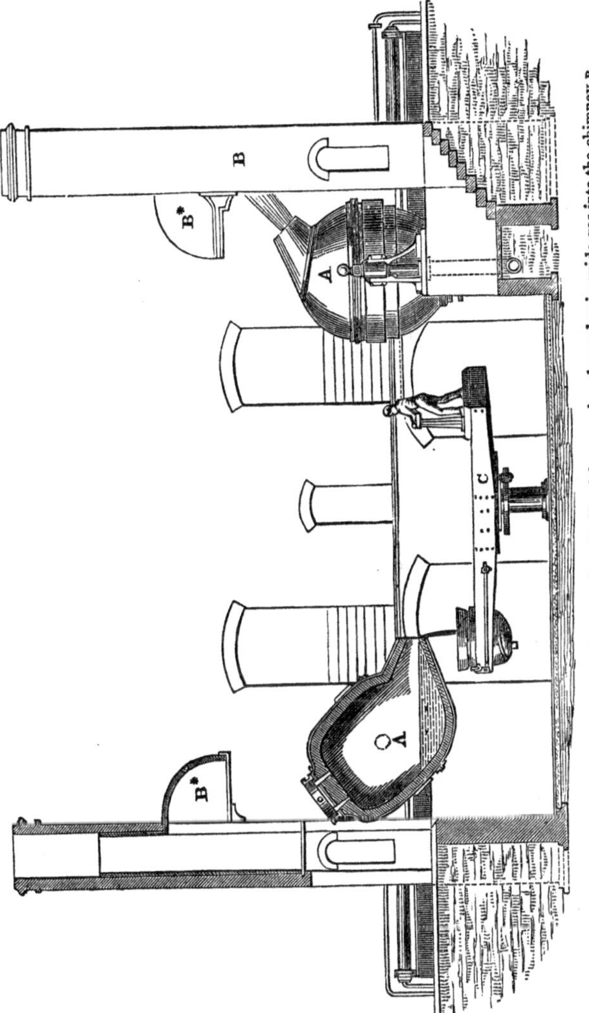

FIG. 24.—Bessemer's process. A, Converting vessel. B*, Hood for carrying the carbonic oxide gas into the chimney B. C, Crane for swinging the ladle under the converter.

is obtained by smelting, in a blast-furnace with charcoal, a spathic iron ore containing a large proportion of manganese, such as that from the Brendon Hills. The result of the analysis of a sample of this material is here given.

Spiegel-eisen.

Iron	82·86
Manganese	10·71
Silicon	1·00
Carbon	4·32
	98·89

The German name *Spiegel-eisen* (mirror-iron) alludes to the brilliant silvery lustre of the metal, the fracture of which exhibits a foliated crystalline appearance of great beauty. The presence of manganese is probably of importance with regard to the use of Spiegel-eisen as an ingredient of Bessemer steel. It is introduced in a melted state, in the proportion of about 1 part to 30 parts of the pig iron employed, into the converter, which is tilted into a horizontal position to receive it, the blast being interrupted during the addition, and afterwards turned on again, for a few seconds, when the converter has resumed its former position, in order to diffuse the Spiegel-eisen through the liquid iron, after which the steel is transferred to the moulds, being poured, for that purpose, into a large iron ladle lined with loam (H, Fig. 25) which is swung under the converter by a crane (G), and, after receiving the metal, is swung back over the wrought iron ingot moulds (K), the steel being run out by raising a fire-clay plug in the bottom of the ladle.

In making a large casting, the ingot mould employed is so massive as to be equal in weight to the metal required to be cast in it, so that it may cool the ingot quickly, and prevent the formation of large crystals in the metal.

Another and more recent form of converter suggested by Bessemer, shown in Figs. 26, 27, has a globular form and seven feet in diameter, the air-blast being introduced through a single tuyère passed through the top of the converter, and

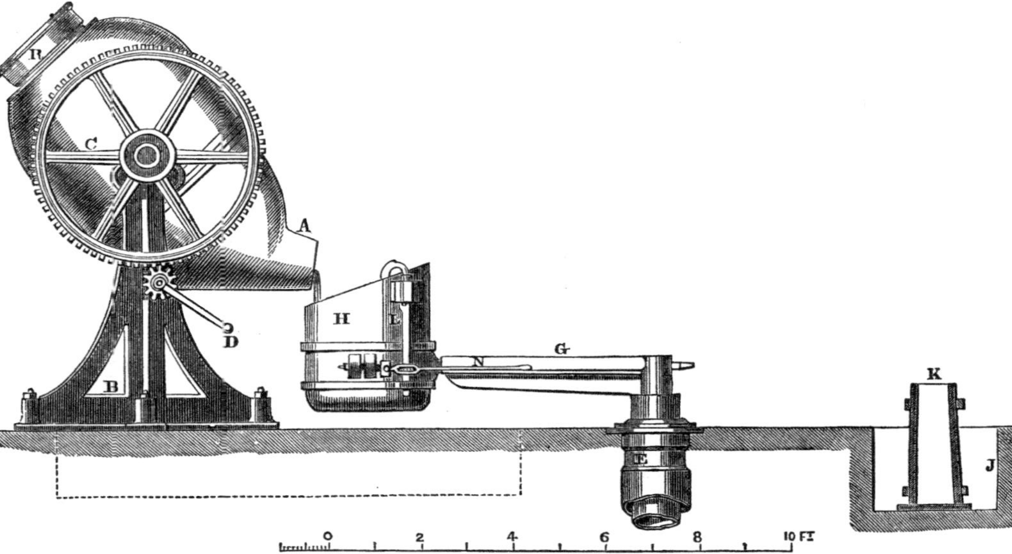

FIG. 25.—Operation of transferring Bessemer Steel from the converter to the ladle. A, The converter. B, Standards supporting the trunnions or axes of the converter. C, Spur-wheel for tipping the converter by means of the pinion moved by the handle D. E G, Hydraulic crane for swinging the casting-ladle H. J, Casting-pit. K, One of the ingot-moulds. L, Rod terminating in a conical fire-clay plug, which closes the opening in the bottom of the casting-ladle, and may be raised by the handle N to permit the flow of the metal into the moulds.

made of circular fire-bricks (D, Fig. 27) strengthened by a stout iron rod passing down the centre, and terminating in a kind of rosette with numerous apertures, through which the air is projected into the liquid iron. When the conversion is finished, the tuyère is lifted out by an ingenious hydraulic crane (E), and the converter tipped by the action of a hydraulic ram, in order to discharge its contents into the casting-ladle. It is said that two such converters are capable

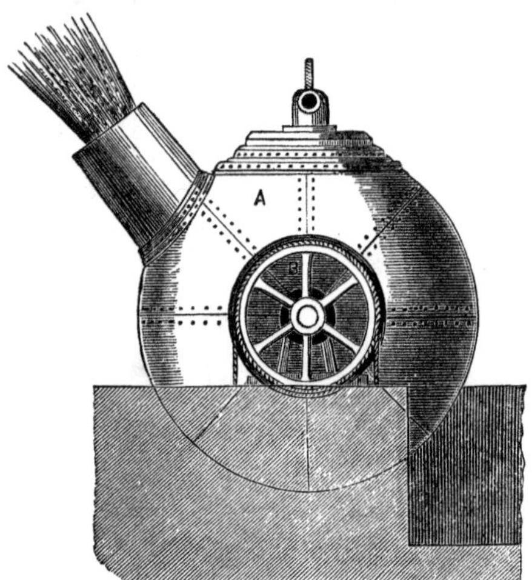

FIG. 26.—Bessemer's globular Converting Vessel.

of producing 200 tons of cast steel weekly, which would require, by the old process of melting blistered steel, 4,730 crucibles, and 760 melting furnaces.

In order to secure perfect uniformity in the composition and therefore in the quality of the Bessemer steel, the practice has been introduced of actually weighing the casting-ladle, running the fused malleable iron from the converter into it, and then weighing it a second time to ascertain precisely the quantity of metal introduced, the calculated proportion of Spiegel-eisen being then added in a melted state, and mixed with the iron by a mechanical agitator made of iron coated with loam, which is made to rotate rapidly in reverse directions for three or four minutes, before tipping the steel into the ingot-moulds. To obtain a sufficient quantity of Spiegel-eisen of uniform composition, the melted metal is run from the blast-furnace on to a revolving table which divides it into drops and scatters them into a cistern

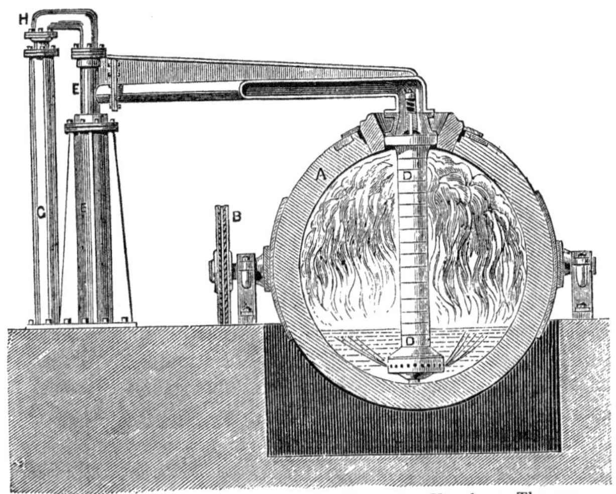

FIG. 27.—Section of Bessemer's globular Converting Vessel. A, The converter. B, Pulley wheel for tipping the converter, connected by a wire rope with a hydraulic ram. G, Pipe conveying the blast. H, Elbow-pipe with telescopic joint.

of water, so that they become converted into granules like shot. By mixing well together, say 500 tons of granulated Spiegel-eisen, so as to obtain a perfectly uniform mixture, it is ensured that each charge added to the iron in the converter will contain the same proportion of manganese, silicon and carbon. The granulated metal is not melted, which would cause an alteration in its composition by the oxidising action of the air, but is merely heated to redness, out of contact with the air, in a kind of crucible, whence it is allowed to drop, through a pipe, into the liquid wrought iron which has been weighed in the casting-ladle.

The extensive manufacture of cast-steel rails by the Bessemer process has led to a very perfect organisation of the works. The cast iron is run direct from the blast-furnace, into a 12-ton ladle mounted on wheels, and taken to the converting-house, where there are six vessels, each capable of converting 5 tons. The 5 tons of steel are run from the casting-ladle into twenty ingot moulds, so that each ingot weighs 5 cwt. These ingots, when removed from the mould, are reheated in a reverberatory furnace, the hearth of which is fixed on a spindle by which it is made to revolve slowly (once in two minutes) so that the flame of the coal fire may act equally upon all the ingots standing separately, on end, and bring them to the proper temperature for the rolling mill. The railway lines of cast steel are far more durable than those laid with puddled bars. These old wrought-iron rails are capable of being converted into steel, by cutting them up, heating them to redness with a little fuel, in the converter itself, and running the melted cast iron in upon them; when the blast is turned on they soon dissolve, and are converted into steel like the rest of the metal.

100 tons of pig iron, treated by the puddling process, yield 75 tons of railway bars, whilst 85 tons of steel bars are obtained by Bessemer's process. The very large steel ingots, sometimes 8 feet long and 3 feet square, and weighing 15 tons, obtainable by this process, cannot be properly forged under the steam-hammer, so that a most ingenious combination of hammer and press worked by hydraulic power has been devised for the purpose.

Cast-steel shot, weighing 300 lbs. each, are made by cutting off pieces from a solid cylinder of steel softened by heat, moulding them by pressure between curved surfaces, and rolling them between two iron tables with corresponding grooves of hemispherical section. The lower table is forced with immense pressure against the upper one by the hydraulic ram, and is at the same time slowly turned on its axis. Three balls are made at once in little more than as many minutes.

The effect of hammering or rolling in augmenting the tensile strength of the cast steel obtained by Bessemer's process is much greater than in the case of malleable iron (See p. 65), for the ingots of Bessemer steel which gave a mean tensile strength of 27¼ tons per square inch, had it increased to 68½ tons by hammering or rolling.

It is alleged by those who are well acquainted with the art of steel-making, that the presence of a minute proportion of silicon in steel is essential to the production of sound ingots, since when this element is entirely absent, the steel disengages gas as it cools in the mould, and boils up with great violence, an effect which is prevented by the addition of ½ part of silicon to 1,000 parts of steel. The silicon present in the Spiegel-eisen employed in Bessemer's process is therefore regarded as of great importance.

A great number of most complete analyses of steel are required to settle this and many other important points in the chemistry of this material, which is daily growing in importance, and with respect to which new theories are continually propounded; at one time *nitrogen* being regarded as an all-important element, at another *titanium*, every theory being apparently supported by a number of chemical analyses and determinations of tensile strength, too often undertaken in the interest of the theory rather than in the unprejudiced search after the truth.

Heaton's Process for the Conversion of Cast Iron into Steel is founded upon the action of nitrate of soda upon the melted metal. The converter is a cylindrical vessel of wrought iron lined with fire-brick, and made to go upon wheels so that it may be easily run under a hood and chimney which fit over it and are attached to it by clamps during the conversion. At the bottom of the converter there is a cavity into which nitrate of soda is rammed, to the amount of one-tenth of the weight of the iron to be converted, and covered with a perforated plate of cast iron. The cast iron having been melted in a separate furnace, is run into the converter through an opening at the side which is then closed by an iron plate. The high temperature of the molten metal decomposes the nitrate of soda, causing it to give off nitrogen and oxygen gases, the latter acting upon the carbon and silicon of the cast iron as in the Bessemer process; but, since the proportion of oxygen to nitrogen is, in this case, nearly ten times as great as in atmospheric air, the action is far more intense, and the conversion of a charge of 15 cwt. is completed in less than 5 minutes. The perforated plate of cast iron melts and mingles with the rest of the metal. Notwithstanding the larger proportion of oxygen in the gas which passes through the metal, the steel is not sufficiently liquefied to be poured out at the conclusion of the process, but is emptied out of the converter in a pasty state, and treated like the blooms taken from the puddling-furnace. It has been claimed for this process, that it consumes less of the iron than Bessemer's process, and that the removal of the sulphur and phosphorus from the cast iron is assisted by the action of the soda derived from the nitrate, so that a good bar iron or steel may be produced from the less expensive descriptions of cast iron which would not yield good results when treated by Bessemer's method.

Nitrate of soda is imported in large quantities from Chili and Peru, where it occurs as an abundant natural product, so that its cost would probably not form a serious obstacle to the general introduction of the Heaton process if the advantages which are claimed for it can be fully realised.

Homogeneous Iron employed for armour-plates and cannon, which has been already alluded to as a *mild steel*, is manufactured by a process which consists in melting a pure description of bar iron (Swedish iron being preferred) with less than one per cent. of charcoal, in the crucibles employed for the manufacture of cast steel, or in a reverberatory gas-furnace.

Siemens devised a process for the production of steel by allowing masses of malleable iron, directly they are reduced from the ore, to dissolve in a bath of melted pig iron heated

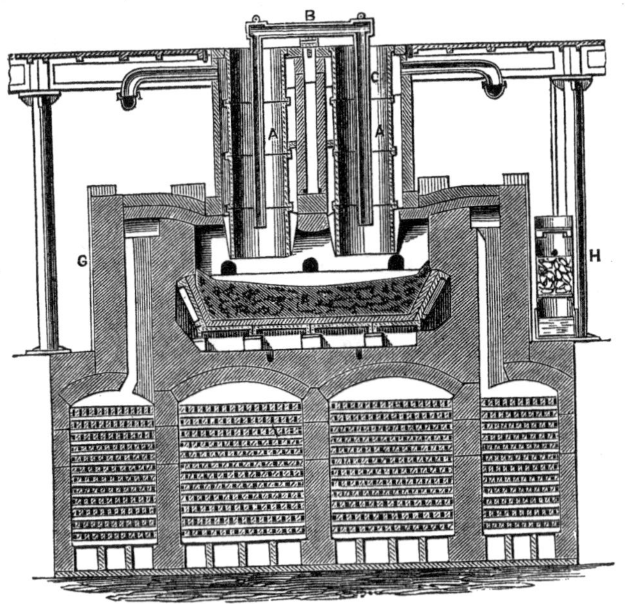

FIG. 28.—Furnace proposed by Siemens, for making cast steel in his Regenerative Furnace. Beneath are shown the chambers containing fire-brick for accumulating the heat of the products of combustion before they pass into the chimney. B, Blast-pipe, from which pipes descend at right angles into the small blast-furnaces A.

in the hearth or combustion-chamber of the regenerative gas-furnace. For this purpose, a small blast-furnace (A, Fig. 28) is constructed above the combustion-chamber, so that its lower opening may rest in the cast iron melted in the latter; this blast-furnace being fed with hæmatite ore and small coke, produces spongy masses of malleable iron, which do not combine with the carbon to form cast iron as in the ordinary blast-furnace, because the temperature is much lower on account of the limited dimensions of the furnace. These spongy masses of iron are speedily dissolved by the cast iron, and the proportion of iron to carbon becomes raised by degrees to that necessary to constitute steel, which is retained in the liquid state by the very high temperature which the regenerative or accumulative principle of the Siemens' furnace renders easy of attainment.

Puddled steel is an inferior description, employed by boiler-makers and ship-builders, and obtained, as its name implies, by arresting the puddling process when there is still enough carbon (from five to ten parts in a thousand) left to constitute a low steel, when the damper is shut, and the puddled balls treated as in the case of iron.

To produce this material, small charges, sometimes only 2 cwt., are puddled, and in bringing the iron *to nature*, the flame of the fire is supplied with less air than when fibrous bar iron is being manufactured, so that less of the carbon may be extracted. The presence of manganese in the iron to be puddled is decidedly favourable to the production of puddled steel, perhaps because the slags containing this metal are more thinly liquid, and cover the surface of the iron more effectually, thus hindering the complete removal of the carbon.

By refining a white cast iron containing manganese in a forge constructed on the same principle as the English refinery hearth (page 50), *natural steel* or *German steel* is obtained. The spathic ores containing manganese yield an iron especially adapted for conversion into natural steel, for which reason such ores are sometimes designated 'steel ores.'

There is as much difference of opinion respecting the effect of the presence of foreign matters upon steel as in the case of bar iron, and for similar reasons, namely, that the quality of steel is so much affected by variations in the mechanical treatment to which it is subjected, that it is difficult to ascertain whether a particular defect in the steel is due to these variations or to the presence of such substances as silicon, sulphur and phosphorus, which are seldom exactly estimated in the analysis of steel by persons whom practical experience has enabled to decide upon the quality of the metal. It is generally allowed that these three elements are injurious to steel, but it is undecided what proportion of each may be present without serious deterioration.

There appears to be little question that steel containing five parts of *silicon* in a thousand does not admit of being forged.

Sulphur confers red-shortness upon steel as it does upon bar iron, but the former appears to suffer less injury than the latter from the presence of a given proportion of sulphur. Steel containing more than two parts of sulphur in a thousand is decidedly brittle at a red heat, so that it is useless for forging and can only be employed for castings, for which purpose it is adapted by its increased fluidity when melted. Red-short steel, like iron having the same defect, appears tougher than other qualities of steel, except at an elevated temperature. Manganese is believed to counteract, to a great extent, the red-shortness caused by the presence of sulphur in steel.

How much *phosphorus* can be tolerated in steel has been made the subject of much discussion. It is said that steel manufacturers object to bar iron as a material for converting into steel when it contains even one part of phosphorus in a thousand.

It is alleged by some that steel made from ores containing titanium is superior in quality, but no conclusive evidence has yet been adduced to show that titanium is really beneficial in its effect upon steel.

A ready test for distinguishing between steel and wrought iron consists in placing a drop of diluted nitric acid (*aqua fortis*) upon a clean surface of the metal, when a greenish-grey stain appears upon the iron, whilst the steel exhibits a black spot due to the separation of carbon.

Hardening, Tempering and Annealing Steel.—When the forging of steel implements is completed, they are nearly as soft and inelastic as malleable iron, and their usefulness depends greatly upon the skill and judgment with which the subsequent operations upon them are conducted. The soft steel is converted into hard steel by being plunged, when red-hot, into water, or sometimes into oil. After this process of hardening, the steel is found to have increased slightly (about one-fiftieth) in volume, and whereas a part of the carbon in the soft steel appears to exist in an uncombined form, and is left undissolved when the metal is acted on by acids, the whole of the carbon in the hard steel is in a state of combination, so that the effect of chilling upon soft steel, in converting it into hard steel, is analogous to that upon grey cast iron in converting it into white iron.

The hardened steel is very brittle, and not unfrequently cracks spontaneously like unannealed glass. A partial explanation of this is afforded by the increase of volume which attends the hardening, for when the outer layer of particles is chilled, those beneath, which are still in a soft state, and comparatively free to move, are restrained from expanding to the proper volume of hardened steel, and the mass is in a state of unnatural strain or tension. Steel which has been hardened in oil instead of water, is tougher and less brittle, which may perhaps be accounted for by the lower specific heat of oil, in consequence of which it abstracts heat less rapidly from the steel, and allows more time for the particles to acquire their proper position in the mass. It is sometimes preferred to heat the oil to about the boiling-point of water before immersing the red-hot steel, in order that the cooling may be still less sudden. Saws are always hardened in oil, for if water be employed they become bent and twisted.

The temperature to which the steel should be raised, in order that it may acquire the highest degree of toughness when plunged in oil, varies for every description of steel, and is usually determined, for each ingot, by a preliminary experiment upon a small sample which is afterwards broken in the testing machine. In toughening the steel tubes for lining wrought iron guns, they are heated to the approved temperature (by a wood fire to avoid contamination with sulphur from coal) and plunged into a bath containing several hundred gallons of rape-oil kept cool by an outer cistern through which cold water flows. They are allowed to cool in the bath for 12 hours or more.

If hardened steel be again heated and allowed to cool slowly, so that there may be less difference between the rates of cooling of the inside and the outside of the mass, the tension or strain of its particles is reduced, and it becomes far less brittle. In the case of a large casting such as those made by Krupp, the chilling effect of the air alone would suffice to render the steel too brittle to be forged, and it is found necessary to delay the cooling of the outside of the mass by keeping it surrounded with hot cinders, so that a casting of sixteen tons will require about three months to cool down.

The following results, quoted from Kirkaldy's 'Experiments on Wrought Iron and Steel,' illustrate in a very striking manner the effect of different modes of cooling upon the tensile strength of steel, the bars being fixed at one extremity and stretched by the action of a weighted lever attached to the other, until they snapped.

	Breaking weight in tons per square inch	Elongation per cent.	Character of fracture
1. Highly heated and cooled *in water*	30	0	Entirely granular
2. Highly heated and cooled *slowly*	36½	22	Entirely fibrous
3. Moderately heated and cooled *in oil*	53	14½	⅓rd granular; ⅔rds fibrous
4. Highly heated and cooled *in oil*	58	2½	Almost entirely granular

From these experiments (which are corroborated by others), it appears that the tensile strength of steel hardened in water, approaches most nearly to that of the best bar iron, although its fracture is so widely different, while that which has been hardened in oil exhibits the highest tensile strength, with a fracture scarcely differing from that of the weakest sample, although it supported a breaking weight of nearly twice the amount.

The relation between ductility and fibrous structure is also well exhibited here; for the bar which stretched $\frac{22}{100}$ths of its original length before breaking was entirely fibrous in its fracture, whilst that which stretched only $\frac{2}{3}$rds of that amount was only $\frac{2}{3}$rds fibrous, and that which did not stretch at all showed no fibre whatever.

In proportion as the hardness of the steel is reduced by this process of annealing or *tempering*, its flexibility is increased, which allows of the production of steel implements of various degrees of flexibility adapted to their different uses. If a knife-blade, for example, be made red-hot, it will be found, after cooling, capable of being easily and permanently bent, as if it were made of wrought iron; its temper is then said to be spoilt, so that a careful regulation of the temperature is required in letting the hardened steel down to the proper temper. Few processes are conducted less according to definite rules than the tempering of steel, the experience of the workmen being more relied upon. In some cases the steel implements, having been raised to a certain tempering-heat, are allowed to cool in air, or more slowly, in sand or charcoal-powder; in others, the workman *fixes or clinches the temper* by chilling in water as soon as the proper degree of softness has been attained. In every case it is important that the steel be raised to a definite degree of heat before the cooling process is commenced. A very general method of fixing this temperature is to watch the colour which a portion of the steel, polished for the purpose, assumes, in consequence of the formation of a very thin film of oxide of iron upon its surface, by the action of the oxygen of the air. The oxide which forms upon the surface of heated steel, is really an opaque black substance, but very thin films of it are capable of decomposing light so as to exhibit a colour varying with their thickness, exactly as the film composing a soap-bubble, though really colourless, exhibits colours variegated according to the thickness of its different parts.

Thus, when steel is heated to about 430° F., an extremely thin film of oxide of iron is formed upon its surface, causing it to assume a faint yellow colour; at about 450° the film is slightly thicker, and shows a pale straw yellow; a full yellow colour appears at about 470°, becoming a brown yellow at 490°, which becomes brown, variegated with purple spots, at 510°; when the temperature has reached 530°, the entire surface has become purple, which gives place to a bright blue at 550°, a full blue at 560°, and a very dark blue at 600°; after which the film of oxide of iron becomes so thick as to absorb all the light which falls upon it and to appear black.

Lancets, which must be very hard in order that they may be ground to a keen edge, are tempered to the faint yellow tinge, whilst razors and surgical knives, which must be less easily broken, are tempered to the straw yellow. Penknives are tempered upon an iron plate over the fire, the blades being laid upon it on their backs until they have acquired the full yellow colour. Cold chisels and large shears for cutting iron must stand rougher usage, and are therefore tempered to a brown yellow, whilst the brown with purple spots marks the tempering heat for axes and plane-irons. Table knives are heated till they acquire a purple colour in order to let them down to the proper temper, and articles in which great elasticity is required, such as swords and watch-springs, are tempered to a bright blue, while saws are brought to the highest tempering heat, at which the dark blue colour shows itself. This temperature, of about 600° F., is that at which oil boils and inflames, so that a bath of oil is very frequently used in tempering, the articles being immersed in it, and the temperature ascertained either by a thermometer, or by the volume and colour of the smoke which rises from the oil. Some tools are annealed by plunging them into oil heated to 400° F. and allowing them to cool down in it. Small steel tools, after being hardened by chilling in water, are coated with tallow, heated over a flame till the tallow begins to smoke, and then stuck into cold tallow. Large steel implements are let down to the proper temper by being heated in a kind of oven known as a *muffle.*

In the subsequent processes of grinding and polishing, the tempering colours of steel articles are removed, but they may be seen in watch-springs, which are not so treated. Transparent varnishes of various colours are sometimes employed to protect axes, &c., from rust, and have much the appearance of the tempering colours. Steel articles are sometimes blued, to protect them from rust, by heating them in a sand-bath until a blue film of oxide is formed.

The following table, compiled from Kirkaldy's experiments, before referred to, affords excellent illustrations of the foregoing statements with respect to the hardening and tempering of steel :—

Cast steel for chisels	Tensile strength (toughness) in tons per square inch	Elongation per cent. (flexibility)
Highly heated and cooled in water .	40	0
Same, tempered to yellow . . .	45	0
,, ,, blue . . .	50	0¾
Moderately heated and cooled slowly .	53	7¾
Highly heated and cooled in ashes .	54	7
Low heat and cooled slowly . . .	56	10
,, ,, in coal tar .	63½	8¾
,, ,, in tallow .	64½	7
,, ,, in oil .	73	5
Moderate heat and cooled in coal tar .	74½	6
,, ,, in tallow .	79½	2¾
,, ,, in oil .	82½	2¾
Highly heated and cooled in oil . .	96	3⅓

*Case-hardening**.—This very useful process is employed to convert into steel the external layer of particles of implements made of wrought iron, in order that they may have sufficient hardness to resist friction, possessing at the same

* This term is sometimes erroneously applied to designate the operation of hardening cast-iron or steel by chilling in water.

time the toughness of malleable iron. Keys and gun-locks are among the articles which are so treated. The operation, as sometimes performed, is an imitation, on the small scale, of the process of cementation, for it consists in burying the wrought iron implement in some carbonaceous substance, and raising it to a red heat, when the outer layer of particles acquires enough carbon to convert it into steel. All carbonaceous substances are not equally efficacious for case-hardening; wood charcoal does not answer so well as that obtained from bone or horn; and in some of the old prescriptions urine is an important ingredient. It is an interesting fact that the substance which has of late years been found the most convenient and effectual for case-hardening is the yellow prussiate of potash, a salt in the preparation of which bone, horn, and similar animal substances are indispensable, the carbon which they contain passing into a new form of combination in the salt, whence it is transferred to the iron undergoing case-hardening. The implement to be case-hardened is heated to bright redness, and sprinkled with the finely-powdered prussiate of potash, and as soon as this has been decomposed by the heat, the metal is quenched in water.

The circumstance that those substances are found to be preferable for case-hardening which contain nitrogen as well as carbon, has led many to believe that the former element plays an important part in the production of steel, and attempts have been made by chemists to show that nitrogen is, like carbon, an essential constituent of steel; their success has not, however, been so complete as to induce the general acceptance of a new explanation of the conversion of iron into steel.

Malleable Cast Iron.—This term is applied to the result of a process which is just the reverse of case-hardening, for it consists in removing the carbon, or a large proportion of it, from articles made of cast iron, so as to confer upon them the toughness of malleable iron. It is practised in the case of small articles, such as buckles, which have to be produced in great numbers at a low price. These are made from a superior quality of cast iron, usually from that which has been smelted with charcoal, and are then de-carburetted by being imbedded in some substance capable of imparting oxygen to the carbon at a red heat, and removing it in the form of carbonic acid gas. Powdered hæmatite iron ore (peroxide of iron) is sometimes employed, and sometimes the iron scale or black oxide which is detached in the process of forging bar iron. Manganese ore (peroxide of manganese) also answers the purpose.

Extraction of Malleable Iron directly from the Ore.—The modern method of smelting iron ores in the blast-furnace, so as to obtain cast iron, which is converted by subsequent processes into malleable iron, owes its origin to the necessity created by the great demand for that metal, of extracting it from the poorer ores, such as clay ironstone, which could not be made to yield their iron by a more direct process. In the early history of the metallurgy of iron, there is no mention of cast iron, the intermediate product of the modern iron smelting, the metal being obtained at once in a malleable condition by a process which is still practised, under various modifications, in districts where ores composed of nearly pure oxide or carbonate of iron can be obtained, together with a sufficient supply of the charcoal which is necessary for the operation.

The direct process of extracting malleable iron is commonly spoken of as the *Catalan process*, since it has been practised from a very remote period in the Spanish province of Catalonia, where the magnetic iron ore and hæmatite of the Pyrenees are smelted with the charcoal supplied by the surrounding forests. The smelting works comprise a forge, a blowing machine, and a hammer, but the first alone will be here described in order to illustrate this method of treating iron ores.

The crucible or hearth is a nearly rectangular trough (M, Fig. 29) well built around with masonry, about 17 inches deep, 21 inches long, and $18\frac{1}{2}$ inches wide. The bottom of the crucible is composed of a block of granite, which is supported upon small arches to keep it dry.

That side of the hearth at which the blast from the tuyère (T) enters is perpendicular, being built up of massive pieces of iron (*t*), the blast-pipe, or tuyère, of copper, being supported upon the uppermost piece in such a manner that its inclination to the bottom of the crucible can be varied at pleasure, since this appears to exercise much influence upon the success of the operation. The wall opposite to the blast is built up of wedge-shaped pieces of iron (*s*) and presents a curved surface. The working side of the hearth is composed of three thick pieces of iron placed end to end, the side opposite to it being lined with fire-clay, and having a moderate inclination.

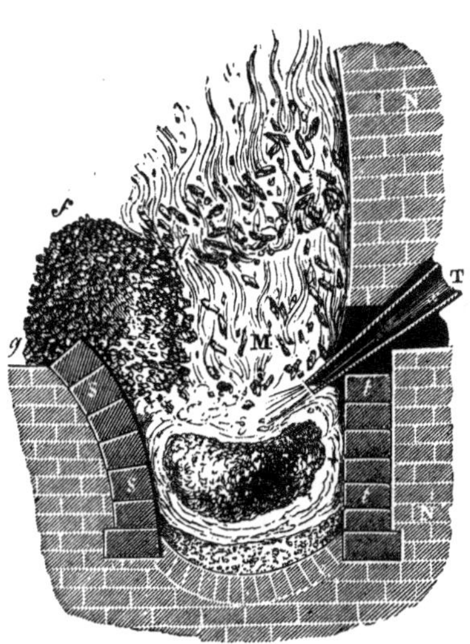

FIG. 29.—Catalan Forge.

To begin the operation of smelting, the hearth is about half filled with burning charcoal, and a shovel is held so as to divide the space above the fuel into two unequal compartments, the larger one, next to the blast-pipe, being filled with charcoal, whilst the other, about half its size, is charged with the ore, previously calcined, broken into small pieces and sifted from the dust; the ore is piled up in a ridge (*f*) upon the side (*g*) of the hearth, so that it may be raked into the fire as the charge sinks down. The shovel forming the temporary partition having been withdrawn, the blast is gradually applied so that it may attain its full force after about two hours, the fuel and ore being continually pressed down into the hearth by the labourers. One portion of the oxide of iron is reduced to the metallic state by the carbonic oxide formed from the carbon of the charcoal and the oxygen of the air-blast (see page 31), but the metallic iron thus produced is not exposed to a sufficiently high temperature to enable it to acquire enough carbon for its conversion into cast iron, and it is obtained in the form of spongy masses of malleable iron or steely iron, according to the proportion of carbon taken up by the metal. This depends, to a great extent, upon the manner in which the operation is conducted. If the *siftings* from the broken ore, moistened with water to prevent their dispersion, be added to the charge in large proportion, the iron contains less carbon and is less steely than when these are employed in smaller quantity, probably because the oxidising action which they exert at a high temperature is unfavourable to the acquisition of carbon by the metal. The iron is also less steely when the blast is directed down to the bottom of the hearth, so that it is less exposed, as it separates from the ore, to the action of unburnt gases very rich in carbon.

A large proportion of the oxide of iron escapes reduction, and combines with the silica contained in the ore and fuel, to form a very fusible silicate of iron, the bulk of which is run off through an opening at the bottom of the crucible.

In about five or six hours, enough ore is reduced to furnish two or three hundredweight of metal, in lumps which are welded together by pressing them with an iron rod, on the end of which they are transported to the hammer, where they are stamped into a compact state, and afterwards forged into bars.

The iron thus obtained is usually of excellent quality, not having become contaminated with foreign matters to the same extent as the melted pig iron from the blast furnace; but the process is a very extravagant one, the ore being made to yield no more than one-third of its weight of metal, with a consumption of more than its own weight of charcoal.

The experiments of Kirkaldy have shown that the quality of iron and steel is pretty correctly indicated by their specific gravities, the specific gravity of rolled steel bars being found to range between 7·8303 and 7·6698; that of rolled iron bars, between 7·7652 and 7·2898, the best qualities having the highest specific gravities.

COPPER.

The use of metallic copper dates from an earlier period than that of iron, although the former metal is by no means so plentifully diffused over the earth's surface as the latter. Copper, however, is of much more frequent occurrence in the pure metallic state, and some of the ores of copper can be much more easily made to yield their metal in a malleable condition.

Native Copper is sometimes found in masses of large size, having a very curious branch-like appearance, each branch being composed of crystals of copper, somewhat deformed, united together. Such masses have been obtained from the southern shore of Lake Superior. Some of the blocks weigh as much as 400 tons, and since the toughness of the metal prevents it from being blasted with gunpowder, much time and labour are expended in cutting the blocks into portable masses with steel chisels. Metallic copper is also found in veins disseminated in granite, in Cornwall and North Wales, and in many other parts of the world. A very remarkable form of native copper is the *copper sand* or *copper barilla* of Chili, which consists of grains of metallic copper mixed with quartz. Native copper, especially that from Lake Superior, is of a very pure description, and is tougher than any but the best specimens of the copper extracted from the ores.

The following table exhibits the composition of the ores of copper:—

Ores of Copper.

	Composition	Copper in 100 parts of pure ore
Red Copper Ore .	Copper, Oxygen	89
Black Oxide .	Copper, Oxygen	80
Copper Glance .	Copper, Sulphur	80
Indigo Copper .	Copper, Sulphur	67
Copper Pyrites .	Copper, Iron, Sulphur	35
Peacock Copper .	Copper, Iron, Sulphur	56
Grey Copper Ore .	Copper, Iron, Sulphur, Antimony, Arsenic	Variable
Malachite . .	Copper, Oxygen, Carbonic Acid, Water	58
Blue Malachite .	Copper, Oxygen, Carbonic Acid, Water	55

Copper Pyrites or *Yellow Copper Ore* is the most abundant English ore of copper, being found in large quantities in Cornwall and Devon. It is also plentiful in Sweden, Saxony, Siberia, and Australia. The colour of pure copper pyrites is a fine brass yellow, but some specimens are much paler, from the presence of iron pyrites. Copper pyrites is much softer than iron pyrites, and the richness of a sample may be in some measure inferred from this character. Although detached specimens of pure copper pyrites may be easily procured, it is always associated, in the vein, not only with the *vein-stone* or *gangue*, generally composed of quartz (silica) or fluor spar (fluoride of calcium), but with arsenical pyrites (composed of arsenic, iron, and sulphur) and tinstone (oxide of tin).

Peacock Ore or *Variegated Copper Ore* is found at St. Austle and Killarney. Like copper pyrites, it is composed of copper, iron, and sulphur, but it contains a larger proportion of copper than that ore.

Grey Copper Ore is one of the most abundant and important ores of this metal, as well as the most complex and variable in composition. Like the preceding ores, it contains the copper in chemical combination with sulphur, but this latter element is also combined with iron, antimony and arsenic, and generally with zinc and silver. The proportion of copper varies between 25 and 40 parts in the hundred, and the silver is very commonly present in sufficient quantity to render its extraction a matter of great importance. Cornwall and Freiberg furnish large supplies of grey copper ore.

Copper-glance, also a Cornish ore of great importance, is a chemical compound of copper and sulphur which is generally free from any important foreign minerals.

Indigo Copper, so named from its dark blue colour, is found in Chili.

Red Copper Ore differs from the preceding ores in being free from sulphur, and since it is found pretty abundantly in Cornwall, Cuba and elsewhere, it plays a prominent part in some of the stages of the process of copper-smelting.

Black Oxide of Copper is found in Chili.

Malachite or *Green Carbonate of Copper* is a very fine green ore, some specimens being so beautifully veined that they are more highly prized for ornamental purposes than as an ore of copper. It is a very pure and valuable ore, but is not abundant in England, being found chiefly in Siberia, the Ural Mountains and Australia.

Blue Malachite or *Azurite* or *Blue Carbonate of Copper* contains a larger proportion of carbonic acid than the green carbonate, and is generally found in the same localities. The mines of Burra Burra in South Australia are noted for malachite ores, which yield copper of excellent quality.

The Cornish copper ores are shipped to Swansea in order to be smelted, the vessels returning to Cornwall laden with the coal required for the tin-works. The ores from Australia, Chili, Cuba (*Cobre ores*), &c., are also received at

Swansea, the neighbourhood of which furnishes an abundance of anthracite coal.

Copper ores are also mined in Anglesea, the Isle of Man, Lancashire, and some parts of Ireland and Scotland.

EXTRACTION OF COPPER FROM ITS ORES.

Probably no other metallurgic operation presents such an appearance of complexity as the smelting of copper ores, but this is due to the great variety of the ores to be treated, which necessitates their introduction at different stages of the process. Thus, a smelting process adapted for copper pyrites must contain provisions for the removal of arsenic and sulphur, which are not present in the carbonates and the oxides of copper, so that the processes of smelting are arranged in such a manner that these ores, as well as the slags obtained in some of the operations, can be introduced after the sulphur and arsenic have been expelled.

In a work like the present, it is not advisable to attempt a detailed account of smelting processes which are subject to frequent alterations in order to suit different lots of ore, particularly when such alterations result from the application of practical experience on the part of the smelter, and do not admit of clear explanation upon simple chemical principles. A general outline only of the extraction of copper from its ores will be given here, and before this is entered upon, it may assist the reader to state that it may be summed up under the following heads:

1. *Roasting processes*, intended to expel arsenic and sulphur, and to convert the iron into oxide of iron.

2. *Melting processes*, intended to remove the oxide of iron by dissolving it with silica at a high temperature, and to obtain the copper as a pure combination of copper with sulphur (sulphide of copper).

3. Roasting and melting, in a single process, to expel the sulphur and obtain metallic copper.

Before being subjected to the first process, the ores are

broken into pieces of the size of a nut, and so assorted that the lot to be smelted may contain about eight or ten parts of metallic copper in the hundred. Moreover, as there is much *gangue* or earthy matter associated with the ores, they are, if possible, so mixed that they may serve as *fluxes* to each other, by producing chemical compounds capable of becoming liquefied by the high temperature of the furnace.

The *fluor* spar*, which is so commonly associated with copper pyrites, derives its name from its power to effect the liquefaction of earthy substances. Fluor-spar is composed

FIG. 30.—Furnace for roasting Copper Ores. BB, Working doors. D, Vault for receiving the roasted ore.

of calcium and fluorine; if it be strongly heated in contact with silica (quartz), which consists of oxygen combined with silicon, the latter takes up the fluorine to form fluoride of silicon gas, whilst the calcium and oxygen unite to produce lime, which combines with another portion of the silica to form a silicate of lime. The silicate of lime would not easily fuse into a slag by itself, but when clay and oxide of iron are present, as is always the case in the melting furnaces, a slag is readily produced.

1*st Process in Copper-smelting. Calcining or Roasting to Expel Arsenic and part of the Sulphur.*—The roasting-furnace

* From the Latin *fluo*, to flow.

or *calciner* (Figs. 30, 31, 32) is a *reverberatory* furnace, with a hearth (A) of large size (about sixteen feet by fourteen) to allow of the ore being spread out in a thin layer upon it. The hearth is commonly built of fire-bricks set on edge and bedded in fire-clay, and the flame is reverberated upon it by an arch of about two feet in average height. At one end of the hearth, near the fire-place, there is an opening or flue (O) through which air may be admitted to the hearth, to furnish the oxygen necessary for the chemical changes effected in the roasting process. On each side of the hearth there are

FIG. 31.—Furnace for roasting Copper Ores. Section through the line X Y of the plan (fig. 32).

two openings (*r*) closed with iron doors, through which the roasted ore is raked out into the arch (U) beneath the furnace. The ore is admitted by opening the hoppers (T) over the arch of the furnace, where it is previously warmed by the waste heat. The fuel employed in the calciners at Swansea is anthracite mixed with one-fourth of bituminous or caking coal, which is necessary to counteract the tendency of anthracite to split up into small pieces and choke the air-passages of the fire, the bituminous coal being softened by the heat, and binding the anthracite together. The fire of the calciners requires special management in order that the ore upon the hearth may be brought to the proper temperature. Anthracite coal is not easily made to burn in an ordinary grate, and, when burning, it raises the bars to so high a temperature that they rapidly oxidise and burn away. To avoid this, a layer of *clinker* or fused ash from the coal is built up on the bars of the grate (F) so as to preserve them from direct contact with the glowing coal, and air-passages are made through this layer, so that the air becomes heated in passing through them, before it actually reaches the fire, the combustion of the anthracite being thus effected by a current of heated air. The oxygen of the air, passing through

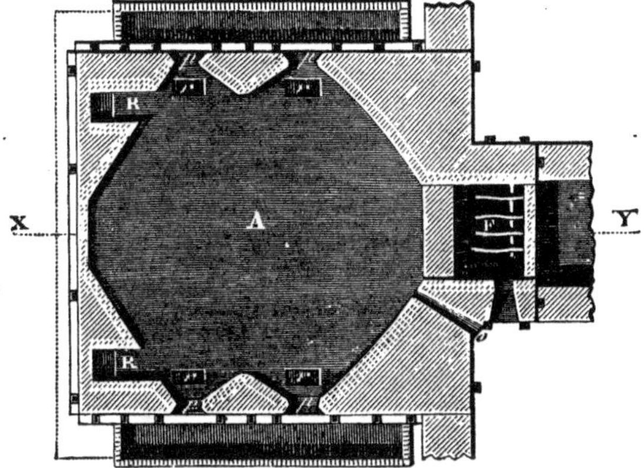

FIG. 32.—Furnace for roasting Copper Ores. Plan at the line Z V of the section (fig. 31).

the column of heated fuel, combines with the carbon to form carbonic oxide (see p. 31), and this gas, being highly heated, takes fire in the air admitted on to the hearth of the furnace, giving a sheet of flame which is drawn through the furnace by the action of the chimney with which the flues (R) communicate, and raises the ore to the temperature necessary for roasting it. Since the air is heavier than the burning gas, a layer of air always exists beneath the latter, separating it from the ore, thus preventing the ore from attaining its melting point, and securing a sufficient supply of oxygen.

Each calciner is charged with three tons of the broken ore, which is spread evenly over the hearth, and roasted for twelve hours, being occasionally raked over through the working-doors (*p*) in order to expose fresh portions to the action of the air, and to prevent any part of the ore from being melted. At this high temperature, the arsenic present in the copper ore combines with oxygen from the air to form *arsenious acid* (white arsenic) which passes, in the form of vapour, into the flues. About half of the sulphur in the ore also combines with oxygen to form *sulphurous acid* gas which passes up the chimney, a small quantity of *sulphuric acid* being also formed and remaining in the ore as sulphate of copper.

Since iron exerts the greater chemical attraction for oxygen, and copper for sulphur, a large proportion of iron acquires oxygen and becomes converted into the *oxide of iron*, while a much smaller proportion of the copper combines with the oxygen from the air to form *suboxide of copper*. When the gases and vapours issuing from the calciners are allowed to escape directly into the air, they form a dense grey cloud of *copper-smoke* which contains the sulphurous acid, mixed with a little vapour of sulphuric acid, the arsenious acid, which condenses in the air to a fine powder, and some hydrofluoric acid gas, produced from the fluor spar. The injurious effect of these products upon the health and vegetation of the neighbourhood has induced the copper smelters to devise means for condensing them by passing them into flues and condensing chambers where they are met by showers of water.

At some works it has been found profitable to convert the sulphurous acid into oil of vitriol instead of allowing it to escape, but in this case it is necessary to prevent the products of combustion of the fuel from mixing with the copper-smoke. *Spence's calciner* employed for this purpose has the fire passing under the hearth instead of over it. This furnace is 50 feet long, and the ore is gradually raked from the cooler to the hotter end as it becomes less fusible. The waste heat of an adjoining smelting furnace is sometimes employed in these calciners, and the calcined ore is raked at once into the smelting furnace. In *Gerstenhöffer's furnace* the ores are crushed between rollers, and allowed to fall over rows of red hot bricks in a vertical furnace, through which a blast of heated air is passed in order to burn the sulphur into sulphurous acid, which is then conducted into the leaden chambers, where it is converted into oil of vitriol.

2nd Process in Copper-smelting. Melting for Coarse Metal, to Dissolve the Oxide of Iron as a Silicate.—It has been seen that the 1st process has had the effect of converting a large proportion of the sulphuret of iron present in the pyrites into oxide of iron, which it is the object of the present process to remove by causing it to combine with silica, to form a compound capable of being melted and separated from the rest of the ore. At this stage the copper ores containing silica (quartz) can be introduced with advantage, provided that they are free from sulphur. It must not be forgotten that, during the process of calcining, a small proportion of the sulphuret of copper in the pyrites has been converted into an oxide of copper, which resembles the oxide of iron in its property of combining with silica at a high temperature, to form a melted silicate which would pass away in the slag, entailing a considerable loss of copper. This is prevented by the sulphuret of iron which is still present in the calcined ore, and, at the high temperature at which the fusion is effected, exchanges constituents with the oxide of copper, forming oxide of iron and sulphuret of copper. The slag from the 4th process, to be presently described, is also appropriately introduced in this fusion, since it contains a considerable quantity of oxide of copper, which exchanges, as above, with the sulphuret of iron in the calcined ore, furnishing more sulphuret of copper to pass into the coarse metal, and oxide of iron to be removed in the slag. The slag from the 4th process (called *metal slag*) also furnishes silica to assist in removing the oxide of iron. In some cases, the smelter adds some fluor spar in order to facilitate the fusion of the charge.

The *ore-furnace* (Figs. 33, 34), as it is called, in which the melting for coarse metal is effected, is also a reverberatory furnace, but its hearth (A) is much smaller than that of the calciner (usually about one-third of the size), because the charge has to be raised to a much higher temperature; for which reason, also, the fire-grate is larger in proportion; the hearth is also slightly inclined on all sides towards a depres-

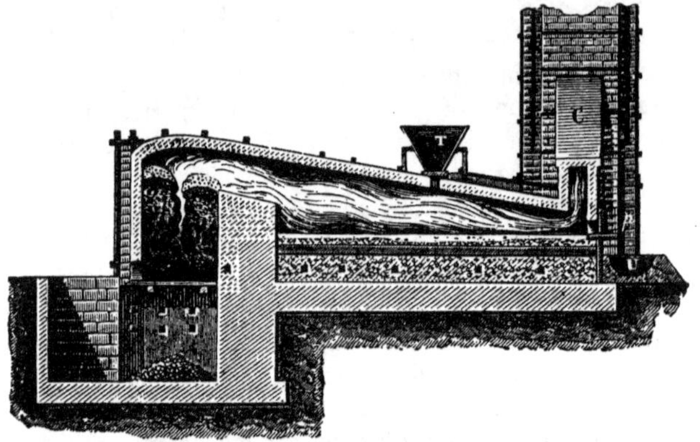

FIG. 33.—Section of Ore-furnace for smelting Copper Ores. T, Hopper for introducing the charge. *p*, Tap-hole for discharging the slag into the slag-moulds U. C, Flue leading to the chimney.

sion or cavity (B) at one side, which serves as a crucible in which the melted coarse-metal collects. The fuel is a mixture of anthracite with one-third of bituminous coal. The charge of this furnace is composed of the following materials, selected for the reasons above given, viz.:—

Calcined or roasted ore, usually about 18 cwt.

Ores containing oxide of copper and silica, 3 cwt.

Metal-slag from process 4, containing oxide of iron, silica, and some oxide of copper, 6 cwt.

Fluor-spar, occasionally.

The slag is the first to fuse, in about half-an-hour after the charge has been introduced, and by degrees the whole of the materials become liquid, and enter into violent ebullition, caused by disengagement of sulphurous acid gas, produced by a secondary decomposition of no importance from a metallurgic point of view, save that the ebullition favours the intimate mixture of the melted matters on the hearth.

After three or four hours, the furnace-man mixes up the melted matters with a rake, and raises the temperature very considerably, to favour the separation of the coarse metal from the slag. In about half-an-hour, the tap-hole (*b*, Fig. 34), which communicates with the cavity in the hearth, is opened, and the *matt** or *regulus* of coarse metal is run out,

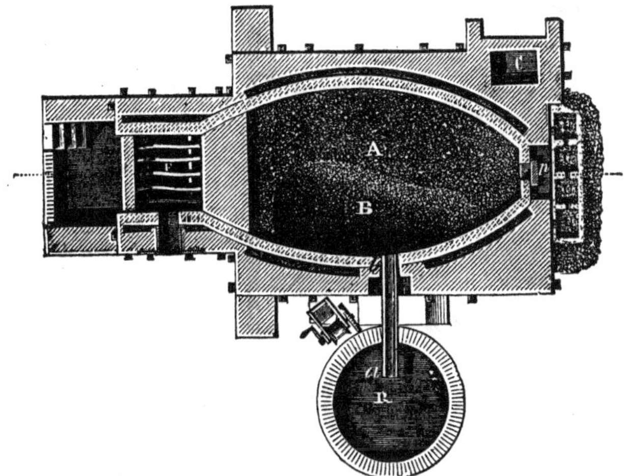

FIG. 34.—Plan of Ore-furnace for smelting Copper Ores. F, The grate. R, Tank for granulating the coarse metal.

through an iron gutter (*a*) into an iron box (G, Fig. 35), perforated at the bottom, and standing in a cistern through which water is constantly running; the coarse metal is thus *granulated* or divided into small irregular grains, in order to fit it for undergoing the next operation.

Sometimes the regulus from two or three operations is allowed to accumulate in the furnace before tapping, the slag alone being raked out before the introduction of a fresh charge.

The iron box containing the regulus is raised from out of

* From the French *mat*, heavy.

the cistern by a winch (W), and its contents are carried to the calcining furnace.

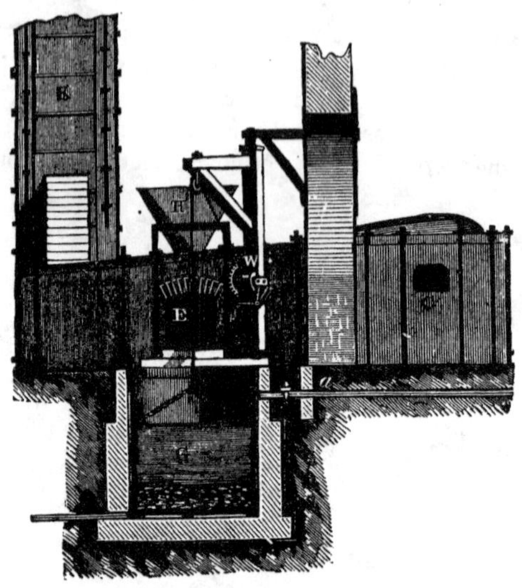

FIG. 35.—Elevation of Ore-furnace for smelting Copper Ores. H, Hopper for introducing the charge. K, Chimney. C, Fire-door. *a*, Pipe for supplying water to the tank.

This coarse metal contains copper, iron, and sulphur in about the same proportion in which they are present in pure copper pyrites, so that the copper amounts to about 33 parts in the hundred, or nearly four times the proportion contained in the raw ore at the commencement of the process.

The slag (*ore-furnace slag*) is raked out into sand-moulds (U, Fig. 34), connected with each other by openings in their sides, where it solidifies into blocks of a black, somewhat glassy, appearance, interspersed with white fragments of quartz. It is used for rough building purposes in the neighbourhood of the copper works. The ore-furnace slag is composed essentially of oxide of iron (ferrous oxide) and silica combined in about equal proportions, and would be spoken of, in chemical language, as a *silicate of iron* or *ferrous silicate.* It contains also a little copper, usually amounting to one part in 140 parts, representing a loss to the smelter which appears unavoidable. Occasionally, a small quantity of regulus is found at the bottom of the blocks of slag, from which it is separated by hand-picking. Fig. 36 exhibits the general arrangements connected with the ore-furnace, and shows the furnace-man discharging the slag.

FIG. 36.—Copper Smelting-furnace.

3rd Process in Copper-smelting. Calcination of the Coarse Metal, to convert more of the Sulphuret of Iron into Oxide.—Now that the earthy matter has been removed in the slag, it is far easier to oxidise the sulphuret of iron than it was in the first calcining process. To effect this, three tons of the granulated coarse metal are roasted in the calcining furnace (Fig. 32) for 24 hours, the temperature being moderated at the commencement, to avoid fusion, and gradually raised in proportion as the removal of the sulphur diminishes the fusibility of the charge, which is raked over every two hours.

About one-half of the sulphur is converted by the oxygen of the air into sulphurous and sulphuric acids, which escape in vapour, another portion of oxygen combining with the iron from which the sulphur has been removed, to form oxide of iron, so that the roasted coarse metal consists essentially of sulphuret of copper, oxide of iron, and some unchanged sulphuret of iron.

4th Process in Copper-smelting. Fusion of the Calcined Coarse Metal to remove all the Iron and to obtain Fine Metal.—The principles involved in this process are the same as in the second process, viz. the conversion of unaltered sulphuret of iron into oxide of iron by exchange with oxide of copper added for that purpose, and the removal of the oxide of iron in the slag, in combination with silica.

The fusion is effected in a furnace which does not differ materially from that employed in the second process, except that there is no cavity in the hearth, which is made to slope from all parts towards the tap-hole (Fig. 34). The charge consists of—

Calcined coarse metal (about one ton)
Roaster-slag from the 5th process } About
Refinery-slag from the 6th process } 12 cwt.
Ores containing oxide and carbonate of copper }

(The roaster and refinery slags contain silica in combination with the oxides of iron and copper.)

These materials are fused together for about six hours, when they divide, as before, into a regulus or matt, and a slag which remains above it. The regulus is sometimes run out into water (like the coarse metal of the second process), when it is called *fine metal*, and sometimes cast into pig-moulds of sand, when it constitutes *blue-metal*, its surface exhibiting a bluish colour, due to its still containing a considerable proportion of sulphuret of iron, in consequence of a deficiency of oxide of copper in the charge. This regulus is composed essentially of copper and sulphur, and contains about 77 parts of copper in the hundred. The presence of a little sulphuret of iron in this regulus gives rise to considerable differences in its colour and appearance, so that it is called by several different names, which do not really imply any important difference in chemical composition.

In some cases, when the charge has contained an excess of oxide of copper, the fine metal has a red brown colour, due to the presence of much red oxide of copper and metallic copper, and a pimply appearance caused by the escape of sulphurous acid gas; it is then called *pimple metal.*

When it is intended to manufacture *best selected copper* for making brass, gun-metal, &c., the blue metal is run into a series of sand-moulds. Since the various impurities which are present tend to collect in a small quantity of metallic copper which is deposited at the bottom of the melted mass, the pigs which are cast first will be the most impure, whilst the others yield the best selected copper in the subsequent operations of smelting. The first pigs yield *bottoms* or *tile-copper* (so called from the shape of the ingots) when smelted.

The composition of a sample of these bottoms is here given, in 100 parts: Copper 74, Tin 14, Antimony 4½, Lead 1, Iron 2½, Sulphur 4. It is evident that the metallic copper which has separated, has decomposed the sulphurets of tin, antimony, &c., contained in the blue metal, and has combined with those metals to form an alloy, which is heavier than the blue metal and sinks to the bottom.

In some smelting-works, where the fine metal is not obtained in so pure a condition, and contains only 60 parts of copper in the hundred, it is again submitted to the two processes of calcining and melting, exactly as in processes 3 and 4, when it yields *black copper* or *coarse copper*, which contains from 70 to 80 parts of copper in the hundred.

The *metal-slag*, as the slag from the 4th process is termed, presents an appearance very different from that of the ore-furnace slag; it is very crystalline and lustrous, and consists chiefly of oxide of iron combined with silica, but it contains a considerable proportion of copper, partly in the form of an oxide in combination with silica, and partly as small particles of metallic copper disseminated through the mass. In some specimens of the metal-slag, the copper appears in very fine brilliant filaments forming *copper-moss.* This slag is usually employed as part of the charge in the 2nd process (melting for coarse metal), but it is sometimes fused in a separate furnace with powdered coal, when a brittle matt is obtained which contains a very large proportion of copper, and is called *white metal,* a name which is also occasionally applied to fine metal.

5th Process in Copper-smelting. Calcining or *Roasting the Fine Metal to remove Sulphur and obtain Blistered Copper.*—The manner in which this process is carried out is varied according to the degree of purity of the fine metal, but the chemical principles which it involves are the following: When a compound of copper with sulphur is heated in air, the sulphur combines with the oxygen of the air, and is thus gradually removed in the form of sulphurous acid gas, the copper also combining with oxygen, and being left as oxide of copper. Further, when an oxide of copper (or compound of copper with oxygen) is melted in contact with a sulphuret of copper (or compound of copper with sulphur), the oxygen of the former combines with the sulphur of the latter to form sulphurous acid gas, and the copper is separated in the metallic state.

The pigs of blue metal are introduced, to the amount of 1½ ton, into a reverberatory furnace, where they are roasted, at a gradually increasing temperature, so as to avoid fusion, for about four hours, in order that a part of the sulphuret of copper may be converted into oxide of copper. When it is judged that this has been effected to a proper extent, the temperature is further raised so as to fuse the materials upon the hearth, the doors of the furnace being closed in order to avoid access of air. As soon as the mass is fairly liquefied, the temperature is somewhat reduced, being again raised towards the close. During this fusion, a violent effervescence is observed in the liquid mass, due to the escape of sulphurous acid gas formed by the union of the sulphur from the sulphuret, with the oxygen from the oxide of copper, whilst metallic copper subsides, in a fused state, and is run out into sand-moulds, where it solidifies into ingots which preserve a blistered appearance caused by the escape of sulphurous acid during solidification.

The duration of the process depends upon the degree of purity of the blue metal under treatment, but it varies between 12 and 24 hours.

A small quantity of slag (called *roaster-slag*) is formed during the fusion, which resembles pumice in its porous texture, but has a dark red-brown colour, and consists of the oxides of iron and copper combined with silica derived partly from the hearth of the furnace, and partly from the sand-moulds in which the ingots of blue metal are cast. This slag contains about 16 parts of copper in a hundred, and is used as a portion of the charge in the 4th process.

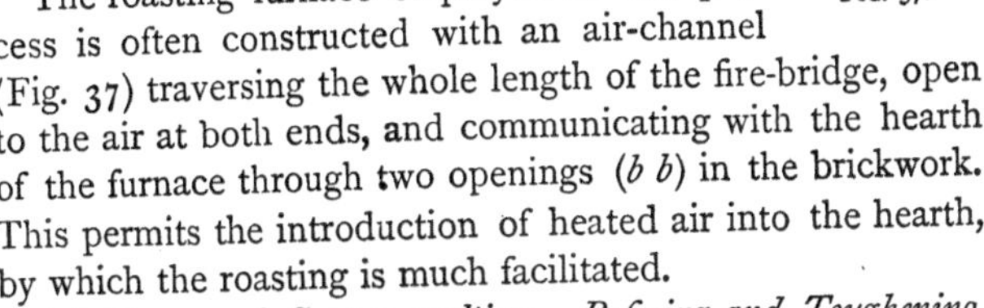

Fig. 37.

The roasting furnace employed in this process is often constructed with an air-channel (Fig. 37) traversing the whole length of the fire-bridge, open to the air at both ends, and communicating with the hearth of the furnace through two openings (*b b*) in the brickwork. This permits the introduction of heated air into the hearth, by which the roasting is much facilitated.

6th Process of Copper-smelting. Refining and Toughening, to purify the Copper.—The pigs of blistered copper are far from pure; it contains considerable proportions of sulphur, arsenic, iron, tin, lead and other foreign substances, varying according to the descriptions of ore employed. In order to remove these impurities, the oxygen of atmospheric air is brought into use. The furnace employed does not differ very materially from the melting furnace used in the 2nd process (Fig. 33), and the blistered copper to be refined is piled, in charges of 6 or 8 tons, upon the hearth, in such a manner as to allow air to circulate freely among the ingots. A moderate heat is applied at first, to allow the oxygen of the air to act upon the blistered copper, an action which is facilitated by the porous structure of the metal. The sulphur then becomes converted into sulphurous acid gas, and the arsenic into arsenious acid which passes off in vapour, whilst the iron, tin, lead and other foreign metals are converted into oxides, as well as a portion of the copper. After being roasted for about six hours, the metal is melted, when a thin layer of slag is formed upon its surface; after raking this off, a small sample of the copper is withdrawn and examined by the refiner, who can judge from the appearance of its fracture how long the subsequent process of toughening will probably occupy. In order to toughen the metal, its surface is covered with wood-charcoal or anthracite, which is renewed from time to time, so as to shield the copper from further oxidation, and the melted metal is stirred with a pole of young wood (usually birch) until a small sample half cut through with a chisel and then broken, exhibits a fine close grain, a silky fracture, and a light red colour; and a small ingot cast for the purpose and hammered when red hot, is found to be soft and free from cracks at the edges. The copper is then said to be at *tough-pitch*, and is taken out in iron ladles lined with clay, and cast into ingots of *tough-cake* copper.

The effect of this process of *poling*, as it is termed, in toughening the copper, depends upon the removal of oxygen from the metal. When the blistered copper has been refined, as above described, by being very slowly melted in contact with air, it is found to have taken up a small proportion of oxygen, which is probably contained in the metal as an oxide (*suboxide*) of copper. The presence of the oxygen, though it does not amount to more than two or three parts in a thousand of copper, has the effect of rendering the copper brittle or *dry*, so that a small ingot of it cracks at the edges when hammered, and its fracture exhibits a deep red colour and a coarse-grained, somewhat crystalline structure. When the melted metal is stirred with the pole, the combustible gases, generated from the wood by the heat, effect the removal of the oxygen from the metal, and bring it by degrees to tough-pitch. If, during the operation of casting the ingots, the surface of the metal on the hearth be not well covered with charcoal or anthracite, the copper will *go back* or become brittle again, in consequence of the absorption of oxygen from the air.

If the process of poling be continued after the copper has been brought to tough-pitch, it becomes even more brittle than before it was poled, an effect which was formerly ascribed to the combination of the copper with a little carbon from the wood; but since analysis has failed to prove the presence of the carbon, the following less simple explanation, based upon experiment, is now generally received. Perfectly pure copper exhibits the malleability and ductility of the metal in the highest perfection, but these qualities are deteriorated by the presence of small proportions of the various foreign matters, such as sulphur, tin, antimony, &c., which cannot be entirely removed in the refining process. The injurious effect of these impurities, however, is counteracted in some measure by the presence of a small proportion of oxygen (not exceeding two parts in a thousand), so that if this element be entirely removed, the copper will be *overpoled*, exhibiting a brittle character due to some of the above-named impurities. On the other hand, if too much oxygen has been left in the metal, the copper is dry or *underpoled.* The effect of overpoling upon the metal may be remedied by allowing air to act for a short time upon the melted copper, so that a small quantity of oxygen may be absorbed by it.

When the copper is intended for rolling into sheets, it is usual to add lead, in the proportion of about five parts to a thousand of copper, just before skimming the surface in order to ladle out the copper. The metal is well stirred after the addition of lead, in order that the action of the air may produce an oxide of lead which combines with the oxides of tin, antimony, and other foreign metals, to form a liquid slag which rises to the surface of the metal and is skimmed off before casting. It is necessary that the removal of the lead from the copper by oxidation should be as complete as possible, since its presence would prevent the scale of oxide of copper from being easily detached from the sheet during the process of rolling, and even $\frac{1}{10}$th part of lead in 100 parts of copper suffices to injure its quality.

This treatment of the metal with lead is called *scorification*, from the *scoria* or slag which forms upon the surface.

The *refinery slag*, skimmed from the surface of the melted copper before commencing the process of poling, has a dull brown-red colour, with a purple shade, and consists almost entirely of an oxide of copper (suboxide) combined with silica derived from the hearth and from the sand-moulds employed to cast the blistered copper. It is employed in the 4th process (fusion for fine metal).

The hearths of the copper-furnaces become strongly impregnated with copper in course of time, and are broken out in order that the metal may be removed from them.

Extraction of Copper from the Bituminous Schists of Mansfeld.—Although most of the copper sent into commerce is extracted by the Welsh process, other methods are sometimes followed on the Continent for the treatment of poor ores, especially when coal is not abundant, for the coal required for the Welsh process amounts to eighteen times the weight of the copper. Thus, at Mansfeld, an ore is extensively worked, which contains not more than four parts of copper in a hundred, in the form of crystals of copper pyrites diffused through a clay slate containing a large proportion of bituminous matter. The consumption of fuel in extracting the copper from this ore is only one-third of that in the Welsh process.

The ore is first roasted in large heaps made up with alternate layers of brush-wood, the bituminous matter also serving as fuel. A heap containing 200 tons of ore will go on burning for fifteen or twenty weeks. In this process, which corresponds to the first calcination in the Welsh method, a part of the sulphur passes off as sulphurous acid, and much of the iron is converted into an oxide.

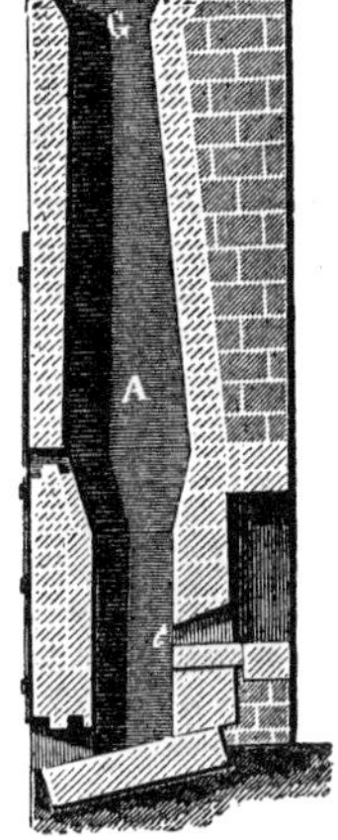

FIG. 38.—Blast-furnace employed for smelting the Bituminous Schists at Mansfeld.

The next process is similar to the Welsh fusion for coarse metal, and consists in melting the roasted ore with some fluor spar, to serve as a flux, some copper-ore containing carbonate of lime, for the same purpose, and some slags containing oxide of copper to decompose the sulphuret of iron and remove the iron as a silicate (page 115). The fusion is not conducted in a reverberatory furnace, as in the Welsh process, but in a small blast furnace (Figs. 38, 39) about 14 feet high, and 3 feet in its greatest width. The blast is supplied by two tuyères (*t*) placed side by side, about 2 feet above the bottom of the furnace, from which the melted matters are conducted through two channels (O O′) into two basins (C C′) about 3 feet in diameter and 20 inches deep, lined with a mixture of clay and charcoal dust; when one of these basins is filled, the channel communicating with it is closed, and the melted matters from the furnace are run into the other basin. The furnace is provided with a chimney (G) 30 or 40 feet high. The fuel employed is either charcoal or a mixture of charcoal and gas-coke, which is charged alternately with the ore, as in an iron blast-furnace. The chemical changes which take place in the furnace resemble those in the Welsh process of melting for coarse metal, and the liquid matter which flows into the

FIG. 39.—Hearth of Blast-furnace employed at Mansfeld.

receiving basins divides into two portions, the lower layer consisting of the sulphurets of copper and iron, and the upper layer of slag composed chiefly of silicate of iron containing but little copper. The slag is ladled out into moulds and employed for building. The *matt*, as the lower layer is called, is removed in crusts, as it solidifies.

If the matt contains less than thirty parts of copper in the hundred, it is again roasted and treated as before, so as to remove more of the sulphuret of iron; but if it contains more than this proportion, it is at once roasted in a special open furnace (Fig. 40), which consists of six separate compartments or stalls with flues running up the back walls in order to create a draught. The matt is placed upon a wood fire in the first compartment, which is then closed by building up a temporary wall; when it has been calcined here for a certain time, it is transferred to the second compartment, and then to the third. It is now introduced into a wooden vessel and washed with water in order to dissolve the sulphate of copper which has been formed by the combination of the oxygen of the air with the sulphuret of copper. The washed matt is roasted again in the fourth, fifth, and sixth compartments, in succession, being treated with water after every roasting. The solution of sulphate of copper thus obtained is evaporated and crystallised, yielding *blue vitriol*, which is sent into commerce.

FIG. 40.—Roasting-stalls employed at Mansfeld.

This operation of roasting, which lasts seven or eight weeks, corresponds to the *calcination of the coarse metal* in the Welsh process.

The roasted matt, containing oxide of iron and sulphuret of copper, is treated as in the *melting for fine metal*, being fused with siliceous slags which dissolve the oxide of iron.

The fusion is effected in a blast furnace similar to that described above, but of smaller dimensions. The liquid matter in the receiving basins divides into three layers, the uppermost consisting of slag, the middle layer of a matt containing 60 or 70 parts of copper in the hundred, combined with sulphur (representing the fine metal of the Welsh copper smelting), which is again roasted and smelted, and the lowest layer of *black copper*, which consists of impure metallic copper, containing about 95 parts of copper in the hundred, with 3 or 4 parts of iron, 1 part of sulphur, and sometimes as much as 120 ounces of silver to the ton. When a sufficient quantity of silver is present to pay for extraction, the black copper is subjected to a process for that purpose, which will be described under silver, and is afterwards refined.

The refining of the black copper, after separating the silver, is conducted in a reverberatory furnace (Figs. 41, 42), the hearth (A) of which is lined with clay and powdered charcoal, upon which the black copper is melted with the flame of a wood fire in the grate (F), and air is thrown upon its surface through two tuyères (*t*), when the oxygen of the air removes the sulphur as sulphurous acid, and converts the foreign metals into oxides which collect as a slag upon the surface. When the refining is nearly completed, a red slag containing much red oxide of copper forms, and a small sample is withdrawn and carefully examined by forging. If it be deemed sufficiently pure, the copper is run out, through the openings (*o*), into receiving basins (B) and removed in *rosettes* by throwing water upon it, and taking off the films of metal thus solidified. The rosettes thus obtained consist of *dry copper*, containing too large a proportion of red oxide. It is refined in the *German hearth* (Figs. 43, 44) which consists of a basin (C) about 16 inches wide, lined with a mixture of clay and powdered charcoal, and furnished with a blast-pipe (T) like that of a blacksmith's forge. The copper being melted in this basin, is covered with charcoal and kept fused until the copper is at tough-pitch in consequence of the reduction of a sufficient proportion of the red oxide ot copper, the same attention and judgment being necessary as in the Welsh process of poling (page 121). The refined metal is then ladled into ingot-moulds.

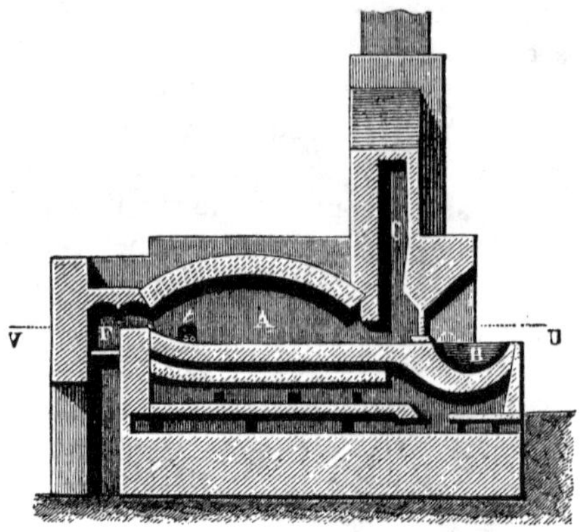

FIG. 41.—Section of Furnace for refining Black Copper, at Mansfeld, made at the line X of the plan, Fig. 42.

When the black copper is not rich enough to be treated for silver, it is at once refined in the German hearth just

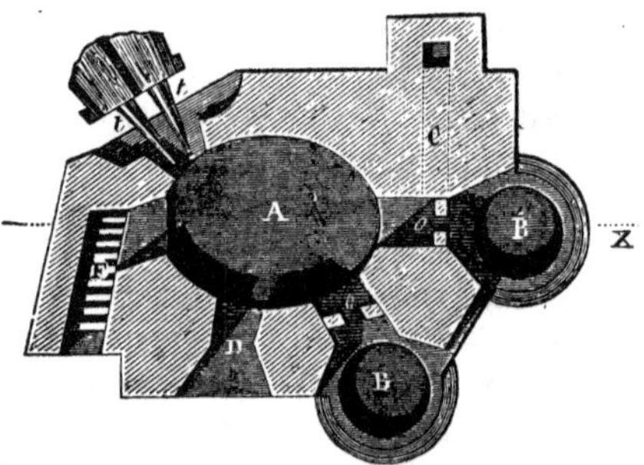

FIG. 42.—Plan of Furnace for refining Black Copper, at Mansfeld, made at the line V U of the section, Fig. 41.

described, but the process is conducted on the principle of the refining of blistered copper in the Welsh process, the impurities being oxidised by the air from the blast-pipe.

The hearth being filled with glowing charcoal, the black copper is placed upon it and gradually fused in the blast, so that the sulphur may burn off as sulphurous acid, and the foreign metals may be converted into oxides which are run off by a channel (*i i'*) provided for them. Alternate charges of black copper and charcoal are added from time to time, and the process continued until the workman perceives, by the inspection of a small sample, that the operation of refining is completed, when the surface is skimmed, and the copper removed in rosettes which are afterwards toughened as described above. The rosettes first removed are sometimes

FIG. 43. FIG. 44.
German Hearth used for refining Rosette Copper, at Mansfeld.

rich in nickel, and are subjected to a special treatment for the extraction of that metal, which is valuable in itself and injurious to the quality of the copper.

Extraction of Copper from Pyrites, on the Tyne.—A considerable quantity of copper is extracted from the residues after roasting pyrites (chiefly Spanish) at the alkali works, for the sake of the sulphur, which is converted into oil of vitriol. These residues contain 2 or 3 parts of copper in 100, with very small quantities of silver. They are mixed with common salt (chloride of sodium), and moderately heated in a reverberatory furnace, when the copper and silver are converted into chlorides of those metals. On washing the roasted mass with water, these chlorides are dissolved, for the chloride of silver, though insoluble in water, dissolves in a solution of common salt. The silver is separated from the solution by adding some iodide of zinc, which precipitates the silver as iodide of silver, and the copper is afterwards obtained by putting in some scrap iron, which combines with the chlorine and causes the separation of metallic copper, to be afterwards melted. The iodide of silver is placed in contact with metallic zinc, which separates the silver, combining with the iodine to form iodide of zinc, which is employed to precipitate the silver from a fresh batch of the material.

VARIOUS DESCRIPTIONS OF COMMERCIAL COPPER.

Anglesea Copper or *Cement Copper* is extracted from the water which is pumped out of the Amlwch mine, near Holyhead, and is called *blue water*, its colour being due to the presence of sulphate of copper produced by the action of the air upon the combination of copper and sulphur present in the ore. This water is pumped into tanks containing scrap iron, which gradually enters into solution as sulphate of iron, the copper being deposited in the metallic state. The fine red colour of the copper, and the pale green of the sulphate of iron, give the contents of the tanks a very beautiful appearance, especially in the sunshine, and the *copper of cementation* thus produced, being almost chemically pure, is of very excellent quality. At Schmöllnitz in Hungary cement copper is also largely prepared in a similar manner.

Rosette Copper or *Rose Copper* is made, chiefly at Chessy in France, by throwing water upon the surface of the melted copper, and removing the solidified metal in films which have a beautiful red coating of an oxide (suboxide) of copper formed by the action of the oxygen of the water upon the metal. These rosettes are plunged into water as soon as they are removed, for if they were allowed to cool in the air, the oxygen would convert the red oxide into a black oxide, which would spoil the colour of the copper.

Japan Copper resembles the preceding, in colour, and is cast into ingots weighing only six ounces each, for exportation to the East Indies. It is coloured by being thrown into water as soon as the ingot has solidified.

Copper is sometimes cast into thin plates by pouring into

the mould enough metal to form a single plate, which is allowed to cool before pouring in a fresh quantity, when a film of suboxide of copper is formed upon the surface of the first plate, which prevents it from adhering to the next, so that the plates are easily separated when the moulding-case is removed.

Bean-shot Copper is made by pouring melted copper through a perforated ladle into a vessel of hot water, when it forms round fragments like shot, which are very convenient for the manufacture of brass. When cold water is used, the metal is obtained in flakes, which are termed *feathered shot.*

In order to remove the scale of oxide from rolled copper plates, before sending them into the market, the somewhat inexplicable course is adopted of soaking them for a few days in urine, then heating them in a reverberatory furnace, rubbing them with a piece of wood, and plunging them when hot into water, when the scale detaches itself. The sheets are then smoothed between rollers.

Effect of the presence of Foreign Matters upon the quality of Copper.—From the circumstance that the refiner tests the quality of copper by forging a *hot* sample, it will be inferred that the effect of impurities upon its malleability and tenacity is more perceptible at a high than at a low temperature. The foreign matters which commercial copper is liable to contain are arsenic, sulphur, antimony, tin, bismuth, lead, silver, iron, and nickel. Of these, sulphur and antimony are generally considered the most injurious in diminishing the malleability and tenacity of the metal. Arsenic is very commonly found in copper, amounting, in some of the Spanish coppers, to as much as one part in a thousand, and was formerly supposed to be as injurious to the quality of the copper as antimony is, but modern experience has shown that copper may be easily rolled and drawn into wire even when it contains a considerable proportion of arsenic. A small proportion of tin is believed to increase the toughness of copper, but bismuth and nickel have the opposite effect.

The conducting power of copper for electricity is reduced in a most striking manner by the presence of foreign matters, so that, in the construction of telegraphic apparatus, it is important that the purest attainable copper wire should be employed. Pure copper is scarcely inferior to silver (see p. 9) in its conducting power, and the conducting power of the native copper from Lake Superior, which is almost pure, stands to that of pure copper in the proportion of 93 to 100, whilst the Australian (Burra Burra) copper, also very pure, has a conducting power of 89, and the Spanish copper, which contains much arsenic, has a conducting power only one-seventh of that of pure copper, or in the proportion of 14 to 100.

The addition of a small proportion of phosphorus (about five parts in a thousand) to copper is found to harden it and somewhat to increase its tenacity; it is also said to render it less liable to corrosion when exposed to the action of sea-water.

By adding arsenic to copper, in about the proportion of one to ten, a white somewhat malleable metal is obtained, which is not easily tarnished by air, and is much harder than copper. This compound, which is employed for clock dials and for thermometer and barometer scales, is made by heating five parts of copper clippings with two parts of white arsenic (arsenious acid) arranged in alternate layers and covered with common salt, in a covered earthen crucible.

TIN.

This metal is scarcely (if ever) found in the metallic state, but is extracted from the ore known as *tinstone,* which is an oxide of tin, or combination of the metal with oxygen. Cornwall has been noted for its tin mines from a very remote period; tinstone is also found in Bohemia, Saxony, Malacca and Banca, the *straits tin* obtained from the last-named localities being much valued on account of its purity. Siberia, Sweden, North and South America, and Australia also furnish tin ore, though in smaller quantity. Tinstone is found either as *stream tin ore* or *mine tin ore.* The former is also called *alluvial* * *tin ore* from its occurrence in the mineral matter deposited by torrents in the valleys adjacent to the veins of mine tin ore, and is much purer than the latter, because it has been mechanically separated, by the action of the stream, from the foreign minerals which were associated with it in the vein. Occasionally, it is found in well-formed prismatic crystals which are perfectly pure oxide of tin. The mine tin ore occurs in veins traversing rocks of quartz, granite, or clay-slate, where it is associated with arsenical pyrites (see p. 104), copper pyrites, specular iron ore, and a remarkably heavy crystalline mineral called *wolfram* (tungstate of iron and manganese) which consists of tungstic acid (an oxide of the metal *tungsten*) combined with the oxides of iron and manganese. In order to obtain the tinstone in a sufficiently pure state for smelting, the ore is stamped to powder, washed, and calcined.

The processes which are put in operation in order to obtain marketable tin from the raw ore may be summed up under the following heads :

1. Mechanical preparation of the ore.
2. Calcining or roasting.
3. Washing the roasted ore.
4. Smelting.
5. Refining.

1. *Mechanical Preparation of Tin Ores.*—The mine tin ore, as it is raised from the mine, is roughly cleansed from earthy matters by washing it upon a grating under a stream of water. It is then picked over and broken with a mallet, the pieces of copper pyrites being placed aside to be smelted for

* *Alluvio* (Latin), an inundation.

that metal, and the iron and arsenical pyrites rejected. The tin ore is then crushed in the *stamp-chest* (C, Fig. 45) which is a wooden trough lined at the bottom with stamped ore, and provided with a number of massive wooden *stampers* (B) shod with blocks of cast iron weighing about $2\frac{1}{2}$ cwts. These are raised by cams fixed to an axle (A) which is made to revolve by water or steam power, so that each stamper may give about twenty blows in a minute, falling through a space of 8 or 10 inches. These heavy blows

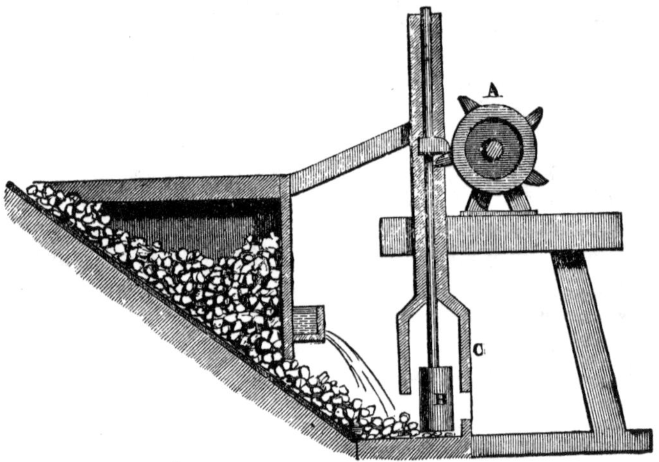

FIG. 45.—Section of Stamping-mill.

speedily reduce the ore to powder, and a stream of water, which constantly flows into the trough, carries this powder through openings in one side of it which are closed with iron plates perforated with about 160 holes in the square inch, so that the larger fragments may not pass through. The holes in the iron plates are conical, having their narrower openings inside the trough, to prevent them from becoming choked. The water carries the powdered ore into a series of reservoirs, in which the ore settles down, whilst the water flows away. Since the tinstone is much heavier than the other substances present in the ore (its specific gravity being

FIG. 46.—Rack for washing Ores.

6·5) the greater part of it is deposited in the first reservoir, the successive deposits becoming poorer as the stream flows on, and the sand, which has a specific gravity of about 2·7,

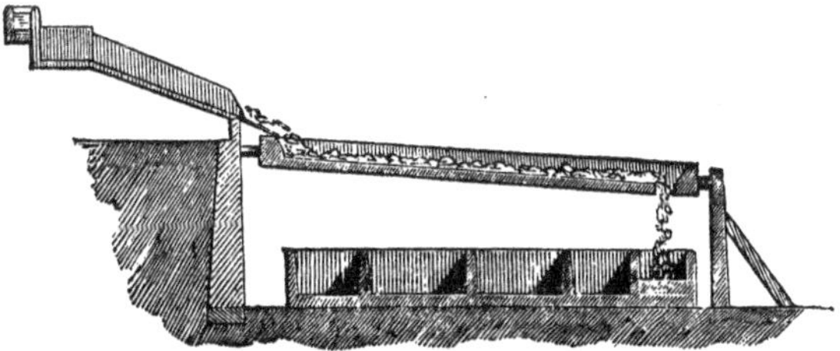

FIG. 47.—Section of Rack for washing Ores.

being in great measure carried away. Various mechanical contrivances are adopted for effecting a further purification of the *slimes* deposited in the reservoirs, in all of which advantage is taken of the high specific gravity of the tinstone. The *rack* (Figs. 46, 47) which may serve as an illustration of these, is an inclined plane of wood with a shallow ledge (*g*), about 9 feet long, having one end 5 or 6 inches higher than the other. It is swung upon a pivot (P) at each end, so that it may be tipped and its contents emptied over the side. About 25 lbs. of the slimes are spread upon an inclined shelf (H), whence they are washed by a stream of water on to the inclined plane (F), when the sand and other earthy portions are carried away by the water, whilst the heavier tinstone, with some pyrites, &c., are left upon the plane; the deposit formed upon the higher portion of the incline is fit for the second process (calcining or roasting), but that formed on the lower part requires another washing. When a sufficient quantity of deposit has been collected on the table, the latter is tipped up sideways, and the upper and lower deposits allowed to fall into separate receptacles.

The *buddle* is a fixed inclined plane worked in a similar manner. The *tossing-tub*, *dolly* or *kieve*, is a tub in which powdered ore is stirred up with water and allowed to settle, the subsidence being hastened by striking the sides of the tub; the lower part of the sediment is of course the richest in the heavier tin ore.

2. *Calcining* or *Roasting the Tin Ore.*—The arsenical pyrites and copper pyrites are too heavy to be entirely removed by stamping and washing, so that the ore is next treated in the *burning-house*, where it is roasted in order to expel the arsenic and sulphur. This is effected in reverberatory furnaces, furnished with horizontal flues several hundred yards long, in which the white arsenic, formed by the arsenic in the pyrites and the oxygen of the air, is deposited in the solid state. From 6 to 8 cwts. of prepared ore is roasted at once, the temperature being very moderate at first, to avoid the fusion of the pyrites, and the ore being frequently raked over to expose fresh surfaces to the oxidising action of the air. The roasting occupies from 12 to 18 hours, and when

Fig. 48.—Brunton's Calciner.

it is completed, nearly the whole of the arsenic has been expelled in the form of vapour of arsenious acid (white arsenic), and much of the sulphur in the pyrites has combined with oxygen, and passed off as sulphurous acid gas; a great portion of the sulphuret of copper in the copper pyrites has also combined with oxygen to form sulphate of copper, a change which is completed by allowing the roasted ore to remain exposed to the air, in a moist state, for some days.

Brunton's Calciner (Fig. 48), which is adopted in some

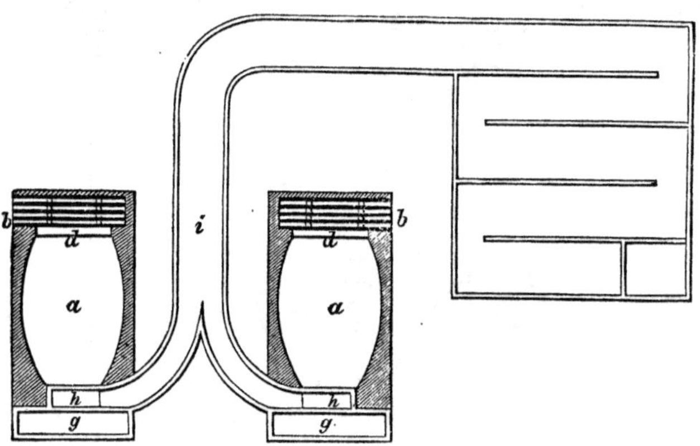

FIG. 49.—Saxon Furnaces for calcining Tin Ore. *a*, Hearth. *b*, Grate. *d*, Fire-bridge. *g*, Chimney. *h*, Flue. *i*, Channel for conveying the fumes into the condensing chambers.

works for roasting tin ores, consists of a circular table (T) of cast iron 8 or 10 feet in diameter, covered with fire-brick, made to revolve at three or four turns in an hour on the hearth of a reverberatory furnace with two grates. The tin ore is allowed to fall from a hopper (H) upon the centre of the table, where it is distributed and turned over by the *spider* (S), an iron frame with projecting arms, which is suspended from the arch of the furnace. By one of these arms the ore which has been gradually carried, by the rotatory motion, to the circumference, is delivered, in a roasted condition, through an aperture (o) under the chimney. In this furnace the roasting operation is carried on without interruption.

In Saxony, the roasting is conducted by a wood fire in reverberatory furnaces (Fig. 49) connected with a flue above 100 feet long, in which the arsenious acid (white arsenic) produced by the oxidation of the arsenical pyrites is deposited. 12 cwts. are roasted in each furnace in 24 hours, yielding 5 or 6 cwts. of white arsenic.

Common salt is sometimes added to the tin ore previously to the roasting, when the chloride of sodium converts part of the arsenic and sulphur into chlorides which pass off in the form of vapour.

3. *Washing the Roasted Tin Ore.*—The next process consists in stirring the roasted ore with water, in a wooden tank, when the sulphate of copper is dissolved by the water, and is drawn off after the ore has settled down, the copper being recovered from the solution by leaving it in contact with iron, as in the case of the blue water of the Anglesea copper-mines (p. 128). By another washing upon the rack, or some similar arrangement, the lighter oxide of iron produced by the roasting of the pyrites is now removed, and the prepared tin ore or *black tin*, containing above 60 parts of tin in the hundred, is ready for smelting.

4. *Smelting the prepared Tin Ore.*—The furnaces (Fig. 50) generally employed in the *smelting-houses* of Cornwall are reverberatory furnaces, having a low arch, with air channels under the fire-bridge and hearth, to prevent injury from the high temperature, the draught being produced by a chimney 40 or 50 feet high. Coal is burnt upon the grate, the flame of which heats the material upon the hearth. About a ton of the prepared ore (oxide of tin) is mixed with one-fifth to one-eighth of its weight of ground anthracite coal, and with a little lime which is intended to flux or liquefy the small quantity of silica still mingled with the ore. Occasionally fluor spar is added with the same object. The mixture is damped with water, to prevent it from dusting, and thrown

upon the hearth of the reverberatory furnace. The doors are closed, but the temperature is kept low at first, for otherwise the oxide of tin would combine with the silica and the lime to form a glass or slag, causing a great loss of tin. By exposing the charge to a gradually increasing temperature during about six hours, the oxide of tin is made to part with its oxygen to the carbon of the coal, yielding metallic tin, which melts and is partly collected in the hollow of the hearth. But as tin is a very light metal (specific gravity 7·3), and the slag inclines to be pasty, it is necessary for the

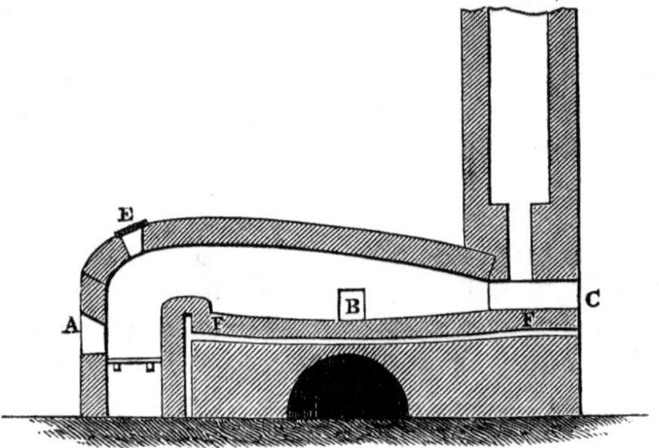

FIG. 50.—Furnace for smelting Tin Ores. A, Fire-door. B, Charging-door. C, Working-door. E, Door for moderating the draught whilst charging the furnace, lest the ore-dust be blown into the flue. F, Air-channel under hearth.

smelter to stir the melted mass in order to facilitate the separation of the tin. The temperature is raised very considerably towards the end of the operation, in order to render the slag as liquid as possible, so that it may not retain too much tin entangled in it. The slag is then raked out of the furnace, and the melted tin is run out into a cast-iron pan, where it is allowed to remain for some time, in order that any slag may rise to the surface, after which it is skimmed and poured into cast-iron ingot moulds.

The slag raked out of the furnace, which is essentially a

combination of silica with oxide of iron, alumina, oxide of tin, and lime, containing as much as $\frac{1}{14}$th of its weight of tin in the form of oxide (beside the metal disseminated through it), is sorted into three kinds. About three-fourths of it, containing very little metallic tin, is thrown away; another portion, with which small globules of tin are intermixed to the amount of about five parts in the hundred, is subjected to the stamping process (p. 132), whilst a third portion, taken from the surface of the melted tin, is so rich in globules of metal that it is worked again in the furnace. About $1\frac{3}{4}$ ton of coal is consumed as fuel for each ton of metallic tin smelted by this furnace.

5. *Refining the Metallic Tin.*—The ingots of tin, as obtained from the smelting-furnace, contain various impurities. Not only are particles of slag, and of the oxide of tin, entangled in the metal, but small quantities of iron, arsenic, copper, sulphur, and tungsten are present in it, and must be removed in order to obtain marketable tin. This is effected by successive operations which are known as *liquation* and *boiling*.

The process of liquation consists in melting out the tin and leaving the impurities behind. The ingots of tin are moderately heated upon the hearth of a reverberatory furnace similar to that employed for smelting the ore, when the bulk of the metal liquefies and is allowed to flow out of the furnace into a *refining-basin*, leaving a residue of impurities upon the hearth. Fresh ingots are introduced from time to time, until about five tons of tin have collected in the refining-basin, which is the case in about an hour after the commencement of the process.

The *boiling* consists in plunging into the tin contained in the refining-basin, which is heated by a separate fire, stakes or logs of wet wood which are held down under the metal by a kind of clamp fixed above the refining-basin.

The heat of the melted metal causes a brisk evolution of steam, which produces all the appearance of boiling in the tin, and carries the entangled impurities up to the surface in a froth which is skimmed off. After about three hours, the wood is taken out, and the melted tin allowed to remain quiet for two hours, when the foreign matters which still remain dissolved in the tin gradually accumulate towards the bottom of the basin, leaving the upper part of the metal nearly pure. (Sometimes *tossing* is substituted for boiling, that is, the tin is well agitated by raising a ladleful of the metal to a considerable height, and pouring it into the bath.) The tin is then ladled out into moulds, either of granite or cast iron, and cast into ingots weighing about 3 cwt. each,* those cast from the upper or purer part of the metal being distinguished as *refined tin*, those from the middle layer as *block tin*,† and those from the bottom being so impure that they must be again subjected to the refining process. The refined tin is very brittle at a temperature somewhat below its melting point, so that when the ingots are heated and allowed to fall from a height, they break up into irregular prismatic fragments which are called *grain tin*. The refiner tests the purity of the tin by casting a small quantity in a stone ingot mould, when the refined tin remains bright and smooth, after cooling; the block tin becomes frosted or crystalline, and the impure tin has a yellowish colour.

The metallic residue left upon the hearth in the process of liquation allows tin of inferior purity to melt out when a stronger heat is applied; this is run out into a small iron basin, allowed to rest, and the upper part of the melted metal ladled into moulds to be afterwards refined: a white brittle alloy containing tin, iron, and other foreign metals is found deposited at the bottom of the basin.

The refined tin will contain 99¾ parts of tin in the hundred, the common or block tin 98½ parts, and the last portion,

* Banca tin is sold in blocks of 40 and 120 lbs. each; Malacca tin in pyramids weighing about 1 lb. each.

† This must not be confounded with *block-tin plate* of which kettles, saucepans, &c., are made, which is not tin, but iron plate covered with a layer of tin.

which requires a second refining, contains only 95 parts. The temperature at which tin is melted before casting is said to be of importance, its malleability being injured if it be cast at too low a temperature.

Reduction of Stream Tin Ore in the Blast-furnace or Blowing-house.—This operation is attended with greater consumption of fuel and loss of tin than that practised in England, but it is largely employed in the tin-works of Saxony. The blast-furnace (A, Fig. 51) is only ten feet high, cylindrical in shape, and surmounted by a conical hood divided into compartments for collecting the dust carried up by the blast, which is forced in by a blast-pipe at *o*, near the bottom of the furnace. The sides and bottom of the furnace are built of granite, the bottom (D) being a single block hollowed out for the reception of the tin which flows out, together with the slag, into a basin of granite (B) lined with a coating of clay and charcoal. If the tin ore contains much oxide of iron, some quartz is employed as a flux; but if much silica is already contained in the ore, lime or finery cinder (p. 52) is employed to form a fusible slag.

FIG. 51.—Blast-furnace or Blowing-house for smelting Tin Ores.

The prepared tin ore and wood charcoal are constantly charged in at the top of the furnace, so as to keep it full, when the carbonic oxide, produced by the combination of the carbon with the oxygen of the air, abstracts the oxygen from the oxide of tin (see p. 31), and the metallic tin runs down to the bottom of the furnace, accompanied by a small quantity of slag formed by the fusion of the silica in the ore with the ashes of the charcoal, and runs out into the *fore-hearth* (B), from which the slag is removed into a tank of water, the tin remaining liquid in the basin. When the latter is full of tin, it is discharged, through a tap-hole, into an iron basin (C), where it is further treated like the tin obtained by liquation (p. 139). The loss of tin in the blowing-houses is twice or three times as great as in the smelting-houses.

At Altenberg, in Saxony, where the ores contain bismuth, they are treated, after roasting, with muriatic acid, for the extraction of that metal. The Saxony tin generally contains a little bismuth.

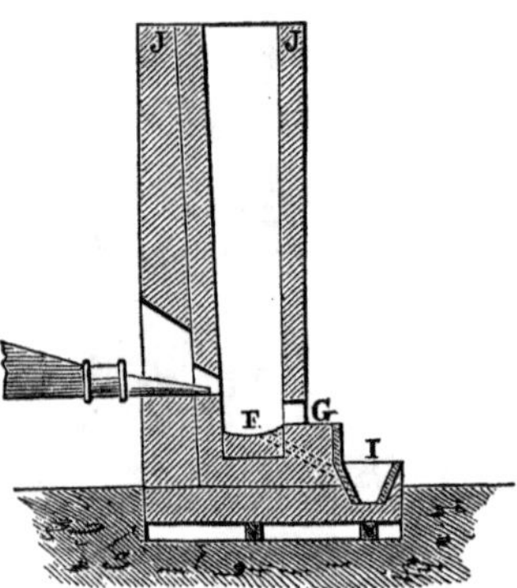

FIG. 52.—Blowing-house at Altenberg. F, Crucible. G, Inclined plane for slag. I, Fore-hearth.

Treatment of Tin Ores containing Wolfram.—Since wolfram is even heavier than tinstone, its specific gravity being 7·5, and that of tinstone 6·5, the tin ore cannot be freed from wolfram by washing. Moreover, the compounds of tungstic acid obtained from wolfram have received, of late years, some important applications in the useful arts; thus tungstate of soda is employed as a mordant by the calico-printer, and is the most effective application for rendering muslin uninflammable; tungstate of baryta is employed as a substitute for white lead. Accordingly, when the prepared tin ore contains any considerable proportion of wolfram, the quantity of this latter is ascertained by chemical analysis, and enough sulphate of soda (salt-cake from the alkali-works) is added to furnish a little more soda than is necessary to form tungstate of soda with the tungstic acid which is present. Some coal dust is added to the mixture, which is then heated on the cast-iron hearth of a reverberatory furnace, when the carbon removes the oxygen from the sulphate of soda, leaving a combination of sulphur with sodium (sulphuret of sodium). A little air is then allowed access to the

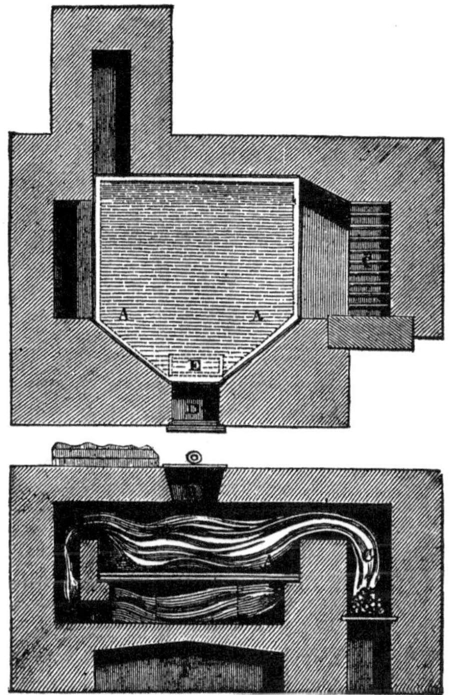

FIG. 53.—Reverberatory Furnace for manufacture of Tungstate of Soda from Tin Ores. B, Opening for introducing the charge. D, Working-door. E, Opening for discharging the contents of the hearth. F, Vault for receiving the finished charge.

heated mass, when its oxygen converts the sulphur into sulphurous acid gas, and the sodium into soda, which combines with the tungstic acid to form tungstate of soda. The furnace (Fig. 53) is so constructed that the flame from the grate (C), after passing over the hearth (A) of the furnace, may return underneath it, so as to heat the charge uniformly. The mass is transferred from the furnace into tanks of water which dissolves the tungstate of soda, to be afterwards obtained in crystals by evaporating the solution. The oxides of iron and manganese derived from the wolfram are still contained in the tin ore, but they are so much lighter that they can easily be separated from it by washing, when the tin ore is ready to be smelted in the usual way.

Tin is remarkable for its property of *creaking* when bent; a bar of the metal, when bent to and fro, emitting a sound as if grains of sand were intermixed with the particles of metal. It has been noticed, in the general consideration of the properties of metals (p. 10), that tin is more easily melted than any other simple metal in common use, and that it is possessed of a high degree of malleability, which is turned to advantage in the manufacture of tin-foil, by rolling and hammering the metal into extremely thin leaves.

Tin is so little acted upon or corroded by air or by weak acids, that it is employed as a coating to protect the surfaces of other metals, such as iron, copper, and brass. Tin-plate is iron covered with a thin coating of tin, and since its manufacture is an important branch of English industry, an outline of it may be here given.

Manufacture of Tin-plate.—The sheet iron employed for the manufacture of the best tin-plate is refined with charcoal (p. 51), though iron refined with coke is sometimes employed, the tin-plate being distinguished accordingly as *charcoal-plate* and *coke-plate.* The last term, however, now usually refers to plate which has been made from puddled iron. A somewhat red-short (p. 69) iron from the Forest of Dean is extensively employed for the purpose, being possessed of great toughness when cold.

In order to obtain iron plates of the required thickness, the bars, $\frac{3}{8}$ths of an inch thick, are cut into pieces 15 inches long and 6 inches wide. Each of these, having been heated to redness, is rolled until its width is increased to 15 inch. It is then again heated, and rolled in the opposite directio. until it is 5½ feet long. The plate is next sprinkled with a little coal-dust to prevent it from sticking together, doubled, again heated, and passed between very smooth rollers until the original length of the doubled plate (2¾ feet) has been increased to five feet. After being again doubled and heated, the four-fold plate is rolled out from 30 inches to 43 inches. It is then cut to the proper dimensions (not exceeding 18 inches by 13), the plates separated, and prepared for tinning. At this stage they are termed *black plates*.

The plates must be rendered perfectly clean and bright, for the slightest impurity would prevent the proper adhesion of the tin.

(1) The plates are bent so that they will stand on end, and arranged in a reverberatory furnace in order that they may be heated to redness.

(2) They are immersed in a mixture of four pounds of muriatic acid with three gallons of water for a few minutes, after which they are

(3) Again heated to redness, when the oxide of iron comes off the surface in scales.

(4) They are hammered straight and passed between rollers of cast iron hardened by chilling (p. 57).

(5) The plates are placed separately, on their edges, in sour bran-water, and occasionally turned, for ten or twelve hours.

(6) They are *pickled*, that is, immersed and stirred about for an hour, in a leaden trough containing diluted sulphuric acid, at a temperature of 90° or 100° F., until they are perfectly bright, the acid having dissolved off all the oxide. The sulphuric acid employed must be free from arsenic, which is apt to produce black spots upon the metal. This operation requires much care and attention on the part of the workmen.

(7) The plates are scoured with sand, under water, and are left under clean water (or sometimes under lime-water), which hinders rusting, until they are required.

(8) The brightened plates are rubbed with bran in order to dry them, and the drying is completed by leaving them for an hour in a cast-iron pot filled with melted tallow.

The process employed in some modern tin-plate works for preparing the plates to be tinned is much simpler than that just described.

(1) They are pickled in warm diluted sulphuric acid for about twenty minutes, which dissolves the black oxide of iron from the surface, the cleansing being completed by scouring them with sand and water.

(2) The plates are annealed by being heated to redness, for twelve hours, in an air-tight cast-iron box capable of containing 1,800 of them, placed in a reverberatory furnace; they are allowed to remain in the box till quite cold, when they are found to have acquired a deep purple colour, from a thin coating of oxide.

(3) They are *cold rolled* between very hard polished rollers, so as to toughen them.

(4) They are annealed as before for about six hours.

(5) The plates are again immersed for ten minutes in warm sulphuric acid, weaker than that employed in process 1, and finally scoured with wet sand.

Tinning.—(1) In order to tin them, they are transferred into another cast-iron pot containing melted tin covered with three or four inches of tallow, and heated over a fire to nearly the inflaming point of the grease. About 340 plates are immersed at once, and are allowed to remain in the tin for an hour and a half, or even longer, according to their thickness, after which the superfluous tin is allowed to drain off by placing them upon an iron grating.

In this operation of tinning, an alloy of iron with tin is formed upon the surface of the plate, and a firm adhesion of the tin coating is thus secured.

(2) The next operation, *washing*, is intended to equalise the coating of tin, and to give the plate a smooth bright

surface. For this purpose, an iron pot with two compartments is employed, both of which are filled with melted grain tin, of high purity, one being designed to receive the superfluous tin from the plate, and the other to give a final coating of pure tin. The plates are immersed in the first compartment, and when sufficiently heated, they are lifted out by tongs, and brushed on each side with a hempen brush; they are then dipped for a moment into the pure tin in the second compartment, which removes the marks of the brush, and afterwards plunged into melted tallow, where they are left for a certain time, and at a particular temperature, in an upright position, when the superfluous tin drains down to the lower edge and forms a rim or *list* which is removed in the next operation. After a certain number of plates have been dipped into the first division of the washing pot, and the tin has become rather impure, about 3 cwt. of it is ladled into the first tinning pot, and replaced by an equal weight of grain tin.

(3) The plates having been placed in another vessel, upon an iron grating, to cool, their lower edges are dipped into the *list-pot*, which contains a layer of melted tin about a quarter of an inch deep; as soon as the list is melted, the plate is lifted and smartly struck with a stick, when the rim of tin drops off.

After the grease has been removed by rubbing the plates with bran, they are ready to be packed in the boxes, which contain 100, 200, or 225, according to the description of plate. About ½ oz. of tin is sufficient to cover a plate of iron measuring 14 inches by 10.

Tin-plate is very durable as long as the coating of tin is perfect, but if any portion of the iron surface beneath should be exposed, it is corroded by rusting even more rapidly than untinned iron plate, because the iron and tin, in presence of the film of moisture containing carbonic acid which is deposited by the atmosphere upon the surface of the plate, form a galvanic pair of which the iron is the metal attacked by the water and acid, so that the plate is speedily eaten through. In the manufacture of the best kind of tin-plate, technically termed *doubles*, and popularly, *block-tin*, a thicker coating of tin is given, and its perfect union and consolidation with the iron plate is ensured by going over the entire surface with a polished hammer upon a polished anvil, an expenditure of manual labour which of course greatly increases the cost of the article.

Tin-plate moiréed.—The beautiful variegated appearance known as the *moiré métallique* is produced by wiping the surface of tin-plate with tow or sponge dipped in a warm mixture of diluted nitric acid with hydrochloric acid, or with common salt, or sal-ammoniac, and well washing with water. The acid liquid dissolves away the smooth surface of the tin and discloses the crystalline structure beneath, the variegated appearance being apparently caused by the reflection of the light in a myriad different directions by the minute faces of the crystals. The moiré may be greatly diversified by heating the plate before applying the acid, and cooling it irregularly by sprinkling water over it, or by directing the blow-pipe flame over its surface before wetting it with the acid. The surface is afterwards covered with a transparent coloured varnish.

Tinning of Copper.—Copper saucepans, stewpans, &c., should always be coated with tin, since most kinds of food are capable of dissolving a little copper, and the poisonous effect of the compounds of this metal becomes perceptible even when very small quantities are present. Fortunately it is much easier to tin copper than iron. The copper surface having been smoothed by rubbing it with a fine sandstone, is made pretty hot, and rubbed over with powdered sal-ammoniac, which has the property of removing the oxide from the surface, and leaving the copper perfectly bright. A little tin is now placed upon the copper, together with some powdered rosin, the latter being intended to form a varnish when it melts upon the surface of copper, which it protects from being oxidised by the air. The copper plate is again heated, and when the tin melts, it is spread with tow over the surface, to which it firmly attaches itself. 200 grains of tin are commonly employed to cover a square foot of copper. The tin employed for this purpose is sometimes alloyed with lead.

ALLOYS OF TIN AND COPPER.

The term *alloy* (from the French *allier*, to blend or unite) is applied by metallurgists to any material which is produced by melting two or more metals together. The properties of the metals which are thus alloyed with each other generally undergo a much greater alteration than can be accounted for, if it be supposed that the alloy is a mere mechanical mixture of its constituent metals, though the proportions in which metals are capable of being alloyed with each other are not so fixed and definite as is the case with substances entering into chemical combination with each other. Since, however, chemists are able to produce, in some cases, very definite chemical combinations of one metal with another, it seems probable that the alloys consist usually of such definite chemical compounds, dissolved in, or mingled with, an excess of one of the constituents over and above the quantity which is actually required to take part in the formation of a chemical compound.

The alloys of tin and copper may be cited in illustration. When two parts of copper are melted together with one part of tin, the mass, after cooling, is found to possess properties very different from those of either of the metals, being very hard, almost as brittle as glass, and having a white silvery fracture. No merely mechanical mixture of two soft malleable metals could produce one which would be hard and brittle, so that the conclusion appears inevitable that a chemical combination has taken place and that a new material has been produced. The tenacity of this alloy is only about one-fifth that of tin, and one-fiftieth that of copper, and it is increased by the addition of either of these metals, which appears to indicate that a further quantity of either of them merely becomes mixed with the alloy. When this alloy is melted with more copper, it is entirely dissolved as long as the metal is liquid, but during the cooling, the alloy tends to separate into two parts, one containing more copper, which solidifies first, and the other much richer in tin. This result would indicate that the original alloy had not formed a true chemical combination with the excess of copper, but had been merely dissolved by it.

The alloy of two parts of copper and one part of tin forms the basis of the *speculum metal* of which the mirrors of reflecting telescopes and other optical instruments are made, arsenic being sometimes added in the proportion of about one-tenth of the weight of the tin, in order to harden the alloy and render it susceptible of a finer polish. A little zinc is often added with the same object. In order to make the alloy, the copper and tin are melted in separate crucibles, and stirred together with a piece of wood. Care is necessary in employing the proper proportions of the metals, for an excess of tin imparts a bluish tinge to the alloy.

Bell-metal is an alloy of copper and tin, the proportions of which are varied according to the size of the bells. Large bells are cast with an alloy of four parts of copper and one part of tin, which is also the composition of the alloy employed for cymbals and gongs. The metal, when first cast, is exceedingly brittle, but it becomes somewhat malleable after being heated to redness and quenched in water. To give it the elasticity which is necessary in order that it may emit a full clear sound, the bell is again heated and allowed to cool slowly. There is an art in the manufacture of a good gong which appears to be possessed by the Chinese alone, and has not yet been successfully imitated in this country.

Gun-metal is composed of 90½ parts of copper and 9½ parts of tin; it is harder and more fusible than copper, and presents great resistance to any strain tending to force its particles asunder. In consequence of the great difference in the specific gravities of the metals, it would scarcely be possible to mix them thoroughly if they were simply introduced together into the reverberatory furnace in which the alloy is prepared. It is customary, therefore, to melt the tin first with twice its weight of copper, so as to obtain *hard metal*, which is then added to the proper proportion of copper melted on the hearth of the reverberatory furnace, care being taken to exclude the oxygen of the air from the hearth, and to mix the metals thoroughly by stirring them with a wooden pole. The formation of the alloy is facilitated by the addition of some old gun-metal. A little more than the necessary proportion of tin is usually added, in order to allow for the unavoidable conversion of a portion of that metal into oxide by the oxygen of the air. When the metals are thoroughly mixed, the oxide is skimmed from the surface, and the gun-metal is tapped into moulds made of loam, the stirring being continued while it is running, to counteract the tendency of the alloy to separate into two parts, as above alluded to. For the same reason, the alloy is run into the mould at a temperature as little as possible above its point of solidification, so that it may not long remain liquid in the mould. In spite of these precautions, a partial separation of the metals always takes place, so that the upper portion of the casting contains more tin than the lower. On this account, it is usual to cast guns with their muzzles upwards, in moulds which are prolonged, in the form of an inverted truncated cone, to about three feet beyond the required casting; the excess of metal, amounting to about twice the weight of the gun, forms what is called a *dead-head*, the weight of which tends to prevent the separation of the metals in the lower part of the casting. This dead-head exhibits a kind of ebullition during the solidification, and a considerable separation, at its upper part, of an alloy containing more than twice as much tin as gun-metal. The dead-heads are cut off before the guns are turned and bored, and are fused in the furnace where the gun-metal is prepared. When the white alloy, rich in tin, separates to any great extent in the lower part of the casting, it produces flaws, or *tin spots*, and the gun is rejected. Comparatively few bronze guns are now cast, since wrought iron and steel have been largely employed for the manufacture of ordnance, being much lighter and stronger.

Bronze has essentially much the same composition as gun-metal, but great variations are made in the proportions of copper and tin, in order to adapt it for special purposes, small quantities of lead and zinc being also occasionally added.

The bronze coinage of this country contains 95 parts of copper, 4 parts of tin, and 1 part of zinc.

The effect of tin in hardening copper was well known to the ancients, who made swords, scythes, nails, &c., of bronze, before the art of working iron and steel had been acquired. The mode of tempering such bronze weapons was just the reverse of that practised with steel, the bronze being rendered soft and somewhat malleable when heated and quenched in water, but hardened again by being heated and cooled slowly.

Tin having a much greater attraction for oxygen than copper has, a considerable proportion of the former metal is lost, as oxide, when bronze is remelted.

The bronze-founder employs a reverberatory furnace so arranged as to prevent access of air to the metal. By the unskilful remelting of old bronze guns, the tin has been reduced to less than half the original proportion.

Britannia-metal.—Tin constitutes the chief part of this and some similar alloys which are employed for the manufacture of spoons, tea-pots, &c., being hardened for such purposes by the addition of antimony and copper. Lead and bismuth are also sometimes added.

The following modifications of bronze are employed for particular purposes:—

	Wheel-boxes or sockets	Stop-cocks and pump-valves	Nails for ships' sheathing
Copper	80	88	87
Tin	18	10	9
Zinc	2	2	4
	100	100	100

ZINC.

Metallic zinc is not met with in nature, and though its combinations with other substances are abundant in certain localities, they are by no means universally diffused over the earth's surface. England is not particularly rich in ores of zinc, and the extraction of the metal is carried out in this country to a very limited extent, most of the zinc required in the arts being imported from Silesia, Belgium and Poland.

The ores of zinc from which the metal is extracted are enumerated in the following table:—

Ores of Zinc.

	Composition	Zinc in 100 parts of pure ore
Blende . . .	Zinc, Sulphur	67
Red zinc ore . .	Zinc, Oxygen	80
Calamine . . .	Zinc, Oxygen, Carbonic Acid	52

Blende derives its name from the German *blenden, to dazzle,* in allusion to its lustre. It usually occurs in black shining crystals which owe their colour to the presence of sulphuret of iron, since the pure compound of zinc with sulphur is white. Blende is also met with of a brown or yellow colour. Black blende is sometimes regarded as a definite compound of sulphuret of zinc and sulphuret of iron, containing 52 parts of zinc in the hundred. The chemical name of blende is sulphide, or sulphuret, of zinc, and the miners often call it *Black Jack.* It is found running in veins through limestone or sandstone, and is commonly associated with galena (sulphuret of lead) and with iron and copper pyrites. Blende occurs in Cornwall, Devonshire, Cumberland, Derbyshire, Ireland, Wales, and the Isle of Man; also at Freiberg, Aix-la-Chapelle, and in North America. It sometimes contains a considerable proportion of cadmium.

Red Zinc Ore is the oxide of zinc, which would be white in the pure state, but is coloured in this ore by the oxides of iron and manganese. It sometimes forms red translucent prismatic crystals, and is found chiefly in New Jersey, in the United States, where it is first smelted for zinc, and afterwards for white pig-iron.

Calamine appears to be so called in allusion to the columnar structure of some specimens of the ore, which gives them some resemblance to a bundle of reeds (*calamus, a reed*). It is a compound of the oxide of zinc with carbonic acid, which would be white if pure, but is usually of a buff or brown colour, due to the presence of oxide of iron, which is objectionable, because it corrodes the clay vessels employed in smelting the ore. Calamine occurs in veins, commonly traversing limestone rocks, and is associated with blende, galena, and *electric calamine*, which resembles calamine in appearance, but becomes electric when heated. The electric calamine is a compound of oxide of zinc, silica, and water (hydrated silicate of zinc), and though it is pretty abundant and rich, it can scarcely be regarded as an ore of zinc, for it does not yield its zinc in the ordinary process for extracting the metal. Calamine is found in Flintshire, the Mendip Hills in Somersetshire, Alston Moor, in Cumberland, at Lead Hills in Scotland, at Aix-la-Chapelle, at Tarnowitz in Silesia, in the north-west of Spain, and in many other places. It sometimes contains more than two parts of cadmium in the hundred. In Spain, the carbonate of zinc is found in combination with the *hydrated oxide* of zinc, so that the ore contains as much as 57 parts of zinc in the hundred. Beds of calamine are reported to have been recently found in Sardinia.

The chief English zinc-works are situated in Birmingham and Bristol, where the ores from the Mendip Hills and Flintshire are smelted; in Sheffield, where the ore is procured from Alston Moor; and at Swansea, Wigan, Llanelly, and Wrexham.

In order to extract zinc from its ores, advantage is taken

of the comparative facility with which the metal is converted into vapour, since it boils and distils freely at a temperature estimated at about 1900° F., a bright red heat, somewhat below the melting point of copper. The ores are calcined so as to obtain the zinc in the form of oxide, which is then mixed with carbon and distilled, when the oxygen passes off in combination with the carbon as carbonic oxide gas, and the zinc is given off in vapour which is condensed again. The mode in which the operation is carried out differs in different works, but the principle of the process is always the same.

Calamine is the principal ore treated in this country, and is sometimes smelted without previous calcination, because the carbonic acid which is combined with the oxide of zinc can be driven off in the smelting process itself; but the calcination or roasting of blende is indispensable, to enable the oxygen of the air to convert the zinc into oxide, and to carry off, in the form of sulphurous acid gas, the sulphur previously in combination with the metal. Care is taken to pick out as much of the galena (sulphuret of lead) as possible, because the oxide of lead which would be formed from it would combine with the silica of the earthern crucibles employed in the smelting process, and would seriously corrode them. The blende is also stamped to powder and washed to free it from earthy matters before calcining.

The ore having been broken into fragments of the size of a nut, the calcination or roasting is effected, as usual, by the flame of a coal fire, in a reverberatory furnace about ten feet long and eight wide, about a ton of ore being spread upon the hearth, and occasionally raked over. The roasting is completed in 10 or 12 hours.

Blende is sometimes subjected to a preliminary roasting in heaps (p. 19), to expel a part of the sulphur before introducing it into the reverberatory furnace.

The roasting-furnace has very commonly two hearths, one above the other, so that the flame, having traversed the surface of one, must pass across the surface of the second hearth before reaching the chimney. The ore is roasted on the upper hearth, which has the lower temperature, for 12 hours, and afterwards, for about the same period, upon the lower hearth. At Moriston, near Swansea, the roasting furnace is about 36 feet long and 9 feet wide, the hearth being divided into three steps, that nearest the fire-bridge being eight inches lower than that near the chimney, and the middle one of intermediate height. 12 cwt. of blende are roasted for eight hours on the coolest hearth near the chimney, then for eight hours on the middle hearth, and finally, for a similar period, upon the hottest hearth near the fire-bridge, fresh charges having been introduced at the other end.

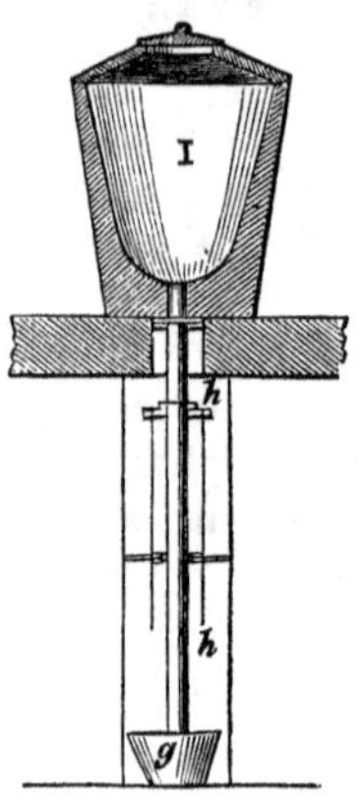

FIG. 54.—Crucible for extraction of Zinc by the English method.

In order to save the clay vessels in which the distillation is effected, those ores which contain much oxide of iron and lime are mixed with others containing clay, which is attacked by those substances, instead of the material of the distilling vessels.

The old English method of effecting the distillation is now almost obsolete, either the Belgian or Silesian process having been adopted in the larger zinc works, but the process is sufficiently interesting to merit a short description.

Old English Method of Extracting Zinc.—The roasted ore is distilled with coke in crucibles (Fig. 54) made of Belgian fire-clay (nearly pure silicate of alumina), furnished at the bottom with iron pipes in which the zinc is condensed. The crucibles are made, generally on the premises, of a mixture of equal parts of fire-clay and old crucibles ground to powder, and each crucible will commonly last about four months, in which period it distils two tons of zinc. The crucibles are 4 feet high and 2½ feet wide, and the iron pipes which pass through the bottom of each crucible are seven inches wide. These pipes are made in two pieces, the shorter length being cemented into the bottom of the crucible, and the other (*h*), about eight feet long, being made to fit on to it when necessary. Six of these crucibles are

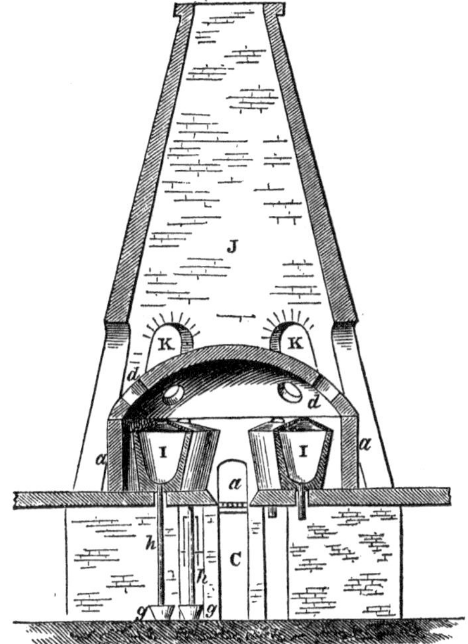

FIG. 55.—Extraction of Zinc by the English method.

usually set in one furnace (Fig. 55), which resembles that used in a glass-house, having a long grate (*a*) running through the centre, on which a brisk coal fire is maintained, the flame from which circulates round the crucibles, and issues through the six apertures (*d*) in the dome (*a*). The external cone (J) serves as a chimney to carry off the smoke, and as a jacket to retain the heat.

The pipe at the bottom having been loosely plugged with a lump of coke to prevent the charge from passing into it, each crucible is charged, through the opening (*d*) above it in the dome, with 4 or 5 cwt. of the calcined ore and 2 or 3 cwt. of broken coke or anthracite. The covers are then cemented on and the fire applied. In a short time, a blue flame appears at the mouth of the tube, beneath the furnace, which is due to the combustion of the carbonic oxide formed by the combination of the carbon with the oxygen of the oxide of zinc. After a few hours, the flame becomes greenish white, from the combustion of the vapour of zinc; it is then extinguished by attaching the longer piece of iron pipe, in which the metal condenses and drops into a vessel (*g*) containing water; the tube (*h*) is removed occasionally in order to clear out the zinc which obstructs it. The distillation occupies about sixty hours, and the ore yields rather more than a third of its weight of zinc, a considerable proportion of the metal being still left in the form of electric calamine (silicate of zinc) which cannot be made to yield up its metal in this process.

The zinc collected in the receivers is, of course, not in the form of a compact mass of metal, because it distilled over in separate drops, and it is intermingled with a considerable quantity of oxide formed as the heated metal dropped through the air into the receiver. It is therefore remelted in an iron pot set over a furnace, and well skimmed, the dross being introduced into the crucibles with a fresh charge. The zinc is cast into flat cakes or ingots, and sent into commerce under the name of *spelter*.*

When cadmium is present in the zinc-ore, it passes over in vapour before the zinc, for cadmium boils at 1580° F., and its presence is indicated by the brown fumes of oxide of cadmium which rise from the flame (*brown blaze*). The metal which then distils over consists of zinc containing a

* The name *spelter* is also given to an alloy of equal weights of zinc and copper employed for soldering brass.

large proportion of cadmium, and is collected separately, the cadmium being employed in making a particular kind of fusible alloy used by dentists, whilst its compound with sulphur forms a yellow paint known as *cadmia*, and other of its combinations are useful to the photographer.

Extraction of Zinc from its Ores by the Belgian Process.— This process is now that most employed in England. The distillation of the mixture of ore and coal is effected in cylinders which are made usually in the zinc-works themselves, from a mixture of raw and hard-baked fire-clay. They are about 8 inches wide and 3 feet long, and are closed at one end. A large number of these cylinders, varying from 40 to 80, are arranged in tiers (*a b c d* Fig. 56), so that they may be heated by the same fire (F), the mouths of the cylinders being placed a little lower than the closed ends. Two, or even four, furnaces are built in the same block, each with two flues running into a common chimney, which is divided into four compartments, each having a separate damper. The furnaces near Swansea are 11 feet wide, $9\frac{1}{2}$ feet high, and 4 feet deep, containing 78 cylinders.

FIG. 56.—Belgian Furnace for distilling Zinc from its Ores.

Coal is burnt on the grate, and its flame penetrates into the furnace through four openings beneath the lowest tier of cylinders. Of course the lower cylinders are thus raised to a higher temperature than the upper rows, so that the former are charged with about 24 lbs. of the mixture of ore and coal, and the latter with 16 lbs. Before the cylinders are arranged in the furnace, the interior of the latter is gradually brought to a high temperature, the front wall being temporarily closed in for that purpose. After four days, the cylinders are set in their places, having been previously heated to redness in a separate furnace.

The mixture of ore and coal, slightly moistened to prevent dusting, is then introduced into each cylinder, to the mouth of which is cemented a conical tube of fire-clay, or of cast iron, about ten inches long. In a short time, the flame of carbonic oxide gas, formed by the carbon with the oxygen from the oxide of zinc, is perceived at the mouth of the tube, and soon afterwards acquires the brilliant greenish-white appearance which indicates that the zinc is beginning to distil over. In order to condense the vapours of zinc, a conical pipe of sheet iron, the smaller opening of which is only two-thirds of an inch in diameter, is now attached. After the lapse of two hours, this is removed and cleared out, the mixture of zinc and oxide of zinc which it contains being worked again with the next charge. The bulk of the zinc has condensed in the cast-iron cones, which are placed in such a position that the melted zinc remains in them, and is raked out by the workmen into a large iron ladle. The sheet-iron cones are again attached, to exclude the air, and after another interval of two hours the condensed zinc is raked out as before. Twelve hours are required to distil the zinc from a single charge of ore, after which a fresh charge is at once introduced, so that the furnaces are kept in continual operation for two months, when they are stopped for repairs. 100 lbs. of the ore furnish, on an average, 31 lbs. of zinc, but a considerable proportion of zinc still remains, combined with silica, in the residue, since the extremely high temperature required to extract it would soften the clay cylinders and cause them to collapse. On account of the high price of coal, clay, and labour, ores containing less than 40 parts of zinc in the hundred cannot be worked in Belgium.

The zinc collected in the iron ladle is skimmed from dross, and cast into ingots weighing from 70 to 80 lbs. each.

At the Vieille-Montagne works, near Aix-la-Chapelle, the ore, a part of which is imported from Sweden, consists of two kinds of calamine, distinguished as *white ore* and *red ore*, the latter containing more iron and less zinc than the former. After being freed from clay by washing, the ore is calcined, either in reverberatory furnaces, or in kilns resembling lime-kilns, in which coal is employed as fuel; it loses about one-fourth of its weight. The ore is then ground to a fine powder, sifted, mixed with half its weight of coal-dust, and distilled in the furnace described above, the white ore, which is richer in zinc, being introduced into the lower rows of cylinders.

Silesian Process for Extracting Zinc from its Ores.—In Silesia, ores which contain only 18 or 20 parts of zinc in the hundred are worked with profit. The calcined calamine is distilled with coal or coke in large *muffles*, or arched ovens (*l*, Fig. 57), 3 or 4 feet long, by 8 inches wide and 18 or 20 inches high. They are made of fire-clay mixed with ground pots, like the cylinders employed at the Vieille-Montagne, but their flat bottoms being well supported throughout their whole length, the muffles will sustain a higher temperature than the cylinders without bending, so that the distillation of the zinc is much more completely effected. The calamine is calcined in reverberatory furnaces, which are sometimes heated by the waste heat of the smelting-furnace. It is broken up into grains about as large as a pea, mixed with about an equal bulk of broken coke or fine cinders, and introduced into the muffle through an opening which is afterwards closed with a fire-clay stopper. The muffles are provided with rectangular earthen tubes (*p*), prolonged by a cast-iron cone and a sheet-iron tube, for the passage and condensation of the vapour of zinc, an opening (*q*) being provided through which the tubes may be cleared from obstruction. Twenty of these muffles are arranged in each furnace, so that the flame may pass well round the top and sides of them, and the firing is very gradual, to avoid all risk of

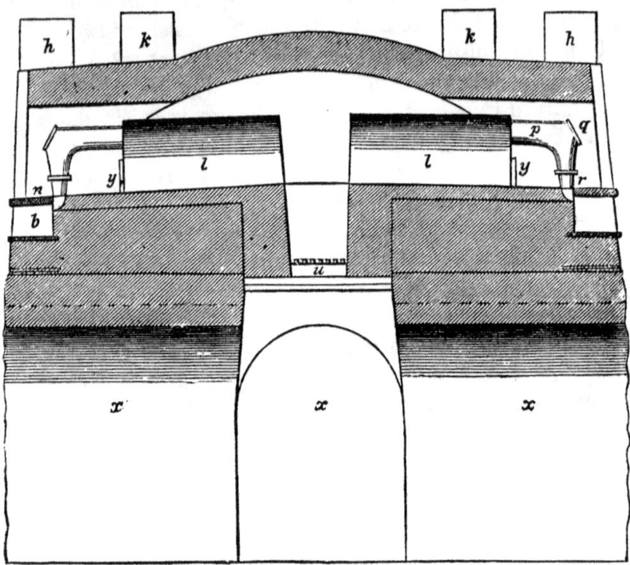

FIG. 57.—Silesian Zinc-furnace; section. *l*, Arched retorts. *y*, Door for removing the exhausted ore. *q*, Door for charging the retorts. *p r*, Tube for condensing the zinc. *b*, Receptacle for condensed zinc. *h k*, Flues. *u*, Fire-grate. *x*, Vaults.

splitting them. The condensed zinc drops from the earthen tube through an opening corresponding to it, in the forehearth of the furnace, and is collected in a small chamber (*b*) beneath. The muffles are charged afresh every 24 hours, and last about 27 days. The residue left in the muffle does not retain more than $\frac{1}{50}$th of its weight of zinc. The zinc is remelted in iron pots lined with clay, since, if the melted metal be in direct contact with the iron vessel, the latter is corroded, and the zinc becomes contaminated with iron, which injures its quality.

To smelt 1 cwt. of zinc from the ore, the coal consumed

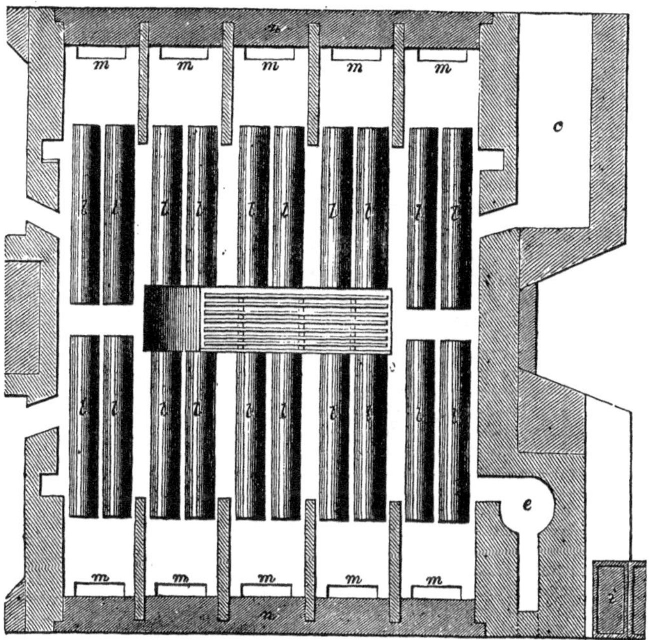

FIG. 58.—Silesian Zinc-furnace; plan. *l*, Arched retorts. *c*, Space for heating the retorts before setting them in the furnace. *e*, Pot for remelting the distilled zinc. *i*, Iron ingot-moulds for casting the zinc.

is, in Belgium, from 6 to 8 cwts., in Silesia from 10 to 15 cwts., and in England 25 cwts.

In the Belgian-Silesian furnaces, as they are called (Fig. 60), the flame is made to pass completely round the muffles (*a*, *b*), and to descend beneath the furnace to the chimney. Thirty-two muffles are commonly arranged in each furnace.

The condensing-tubes (*g*) have a depression at the under side, in which any lead which distils over is deposited, and the zinc runs off in a purer condition, and are sometimes provided with tapping-holes through which the zinc may be run off into ingot-moulds.

At Stolberg, a furnace of this construction has 60 muffles arranged in two tiers, each muffle receiving 70 or 80 lbs. of ore.

In some cases it has been found economical to charge the roasted ore into a cupola or small blast-furnace, in alternate

FIG. 59.—Silesian Furnace for distilling Zinc from its Ores. *i*, Condensing tube. *t*, Openings beneath the hearth for receiving the condensed zinc.

layers with fuel, when the zinc is given off in vapour, which is speedily converted into oxide by the oxygen of the air, and, being collected in condensing-flues, is distilled in the usual way with coal for the production of metallic zinc. The advantages of this process consist in the greater yield of zinc in the distillation, and the absence of the oxides of lead and iron which do so much damage to the clay vessels.

At Bleiberg, the *zinc-dust*, or mixture of finely-divided zinc with oxide of zinc which is first collected during the distillation of the ores, is melted in fire-clay tubes set upright in a furnace; clay pistons attached to iron rods are thrust into the tubes, the pressure causing the finely-divided zinc to run together, so that it may be tapped out from the bottom of the tubes. The metal so obtained is very impure, containing much arsenic and cadmium, which always pass over with the first portions of distilled zinc.

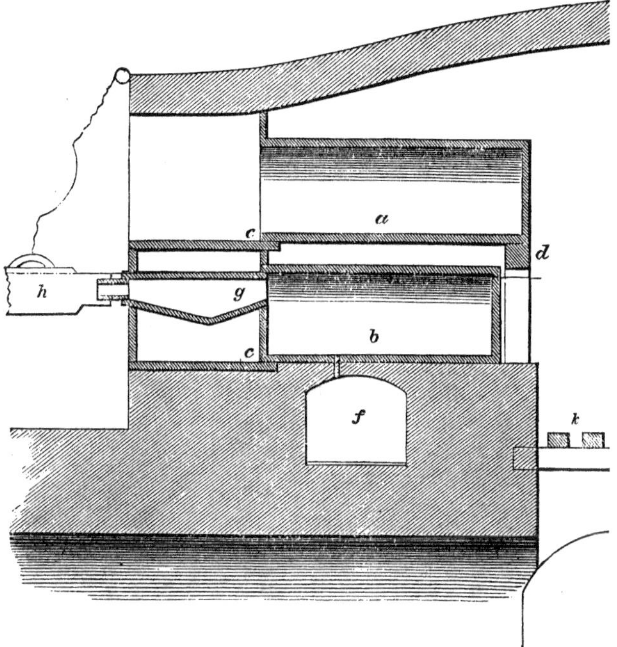

FIG. 60.—Belgian-Silesian Zinc-furnace. *a b*, Arched retorts. *c d*, Their supports. *f*, Flue. *g*, Condensing tube. *h*, Receiver suspended by iron-wire. *k*, Fire-place.

Rolling of Zinc into Sheets.—Before the year 1812, zinc was used almost exclusively for the manufacture of brass, since it is brittle both at the ordinary temperature and at high temperatures, but it was then discovered that a temperature between 200° and 250° F. rendered it malleable and capable of being rolled into thin sheets. For this purpose, however, it is necessary that the zinc should not contain iron or lead, the former of which it acquires when remelted in iron pots, while the lead is carried over in the distillation of the zinc, in consequence of the presence of galena (sulphuret of lead) in the ore.

To prepare the zinc for rolling, the ingots of spelter are remelted upon the fire-clay hearth of a reverberatory furnace, which is made to slope down to a deep cavity or *sump* at one end, into which the melted zinc flows. Lead (specific gravity 11·4) being much heavier than zinc (specific gravity 6·9), and the two metals having very little disposition to alloy with each other, the bulk of the lead settles down to the bottom of the cavity, so that the upper portion of the melted metal is comparatively free from lead. It is then ladled out and cast into ingots of convenient size for rolling, and after these have been again heated to about 250° F. by the waste heat of the furnace, they are passed between cast-iron rollers and reduced to the required degree of thinness.

After being rolled, the zinc always retains a great deal of its malleability, even after cooling, whereas cast zinc may be broken by laying it across an anvil and striking it with a hammer. The fracture of an ingot of zinc presents a very beautiful crystalline appearance and great lustre, and if it contain iron, minute grey spots are visible on the bright faces of the crystals. At temperatures above 400° F. zinc is even more brittle than at the ordinary temperature, and it may be obtained in extremely fine powder by pouring melted zinc into an iron mortar, and well stamping it with the pestle as it solidifies. Powdered zinc has been employed as a metallic paint for protecting iron from rust.

Since the discovery of the malleability of zinc at very high temperatures, this metal has been extensively employed for gutters, rain-pipes, cisterns, baths, chimney-pots and roofing, purposes for which it is eminently fitted by its lightness and by its resistance to the action of air and moisture; for although a bright surface of zinc soon tarnishes, from the formation of a film of oxide of zinc upon it, this film serves to protect the metal beneath from any further corrosion.

An objection to the use of zinc for roofing is the great combustibility of the metal, for, at a red heat, it takes fire and blazes fiercely, producing light white flakes of oxide of zinc, which is used as a paint, under the name of *zinc-white.*

In casting zinc, it is important to avoid employing too high a temperature, not only because zinc may be lost in vapour, but because *burned zinc* is produced; that is, the zinc becomes very much harder, and difficult to cut with a file or chisel, probably because it has dissolved some oxide of zinc.

Galvanised Iron is sheet iron coated with zinc, and is made in a similar way to tin plate, by thoroughly cleansing the surface of the iron, and immersing it in melted zinc which is kept covered with powdered sal-ammoniac (muriate or hydrochlorate of ammonia), this salt having the property of dissolving the oxide of zinc from the surface of the bath. The zinc probably alloys with the iron plate to a slight depth. A small quantity of iron is dissolved by the melted zinc, and a very brittle alloy is deposited in pasty masses at the bottom of the zinc-pot, whence it is removed occasionally with a perforated ladle. This alloy has been found to contain 6 parts of iron and 94 parts of zinc. In some large galvanised iron works, where much of this alloy is obtained, the zinc is recovered by distilling the alloy in clay retorts like those employed in gas-works. A process recently devised for treating this alloy consists in melting it in iron pots, and cooling it slowly, from the bottom upwards, when an alloy of 9½ parts of iron with 90½ parts of zinc is deposited in crystals, which are removed by a perforated ladle, and the liquid zinc is left much purer. The crystallised alloy is distilled, to recover the zinc. The zinc coating adheres more firmly if the iron is previously tinned.

Iron articles are also coated with zinc by connecting them with wires attached to the negative pole of a weak galvanic battery, and immersing them in a solution of sulphate of zinc; this is decomposed by the galvanic current, the zinc being deposited upon the surface of the iron, which is thence said to be *galvanised.* When any portion of the iron surface is exposed in consequence of the abrasion of the zinc, the adjacent portion of the latter metal will protect the iron from corrosion, because the two metals form a galvanic pair of which the zinc only is attacked by the moisture and carbonic acid of the air. It will be remembered that, in the case of tin plate, it is the iron which is the metal attacked.

Galvanised iron is ill-adapted for situations where it is much exposed to the acid vapours sent into the air by some factories, or to the sulphuric acid found among the products of combustion of coal and gas, because zinc is among the metals most easily attacked by the acids. It is often called *corrugated iron,* from the practice of ridging and furrowing the plates in order to strengthen them for building purposes. Vessels of galvanised iron are not well fitted for cooking utensils, since many articles of food are liable to become contaminated with zinc to a hurtful extent.

The low temperature at which zinc is melted (770° F.), and its perfectly liquid condition, recommend it for casting statues, since it runs easily into the very finest lines of the mould. It is only a sixth or an eighth of the price of bronze, and its surface can be coloured so as to resemble that metal.

ALLOYS OF ZINC AND COPPER.

Brass is usually described as an alloy of zinc with twice its weight of copper, and there is some evidence that the metals enter into a true chemical combination in these proportions; but if this be so, the compound is capable of dissolving an excess of either metal, the proportions employed for brass being varied to suit the particular purpose for which it is intended.

Before the year 1780, brass was always made by strongly heating copper in contact with calamine and charcoal or coal. 3 parts of bean-shot copper (p. 129), 3 of calcined calamine, and 2 of charcoal were heated to bright redness in a covered crucible, when the zinc reduced from the calamine by the action of the carbon, instead of passing off in vapour, as in the extraction of zinc from the ore, entered into combination with the copper, producing brass which was found in a melted state at the bottom of the crucible. Brass is still manufactured by this process at Holywell in North Wales.

When metallic zinc is employed for making brass, it is usual to melt the zinc in a crucible, and to add the copper in small portions at a time, until a nearly solid alloy has been produced. This is broken up and remelted with the proper proportion of zinc.

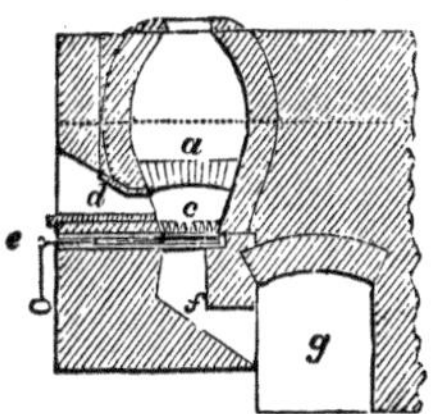

FIG. 61.—Furnace for making Brass. *a*, Arch upon which the crucibles stand. *c*, Grate. *d*, Fire-door. *e*, Damper. *f*, Inclined plane for carrying away cinders into the channel *g*.

The union of the copper and zinc is much facilitated by melting them with old brass. Crucibles made of fire-clay, or of a mixture of fire-clay and plumbago, are employed; convenient dimensions are 16 inches in depth, 9½ inches wide at the mouth, and 6½ inches at the bottom, the sides being 1 inch thick, and the bottom 1½ inch. Eight of these crucibles are heated by a single coal-fire (Fig. 61), the flame of which passes up and circulates around them, and afterwards heats two empty crucibles placed above them, and intended for the subsequent casting of the brass. Suitable proportions of the metals are:

41 lbs. of old brass.
55 lbs. of best selected bean-shot copper.
24 lbs. of zinc.

The crucibles are filled with the pieces of old brass, which are melted down, and leave room for the other metals.

Half of the zinc is then introduced, in small lumps, and covered with coal-dust; then half of the copper and another layer of coal-dust; the rest of the metals is then introduced in the same way, the whole covered with a layer of coal-dust, and the crucibles exposed to the fire for about four hours, when the brass is ready for casting. One of the hot empty crucibles is taken out of the furnace, and placed in another fire so as to keep it red-hot whilst four of the crucibles of brass are emptied into it; the surface is then skimmed, and the brass is poured into moulds made of slabs of granite mounted in an iron frame, the joints being cemented with clay.

The presence of iron in brass is very objectionable, as it gives rise to deficient tenacity and malleability.

The colour of the alloy of zinc and copper is of course dependent upon the proportions in which the metals are employed. Those alloys which contain more than 80 parts of copper in the hundred exhibit a reddish yellow colour, in which the red predominates as the quantity of copper increases, the colour becoming yellow when less than 80 per cent. of copper is present. If the amount of copper be less than 30 parts in the hundred, the alloy is no longer yellow, but approaches more nearly to the colour of zinc.

The various alloys used to imitate gold, before the art of electrogilding was introduced, were all modifications of brass. *Dutch metal* or *Dutch leaf gold*, which is one of the most malleable of alloys, is composed of 11 parts of copper and 2 parts of zinc. It is cast into thin plates between slabs of granite, and rolled into sheets, being occasionally annealed (p. 5). When these are very thin, several are passed through the rolling-press together, and they are eventually cut up, and beaten out to extreme tenuity in piles of 40 and 80, under a hammer worked by water-power, making three or four hundred strokes per minute. *Bronze-powder* (or at least one kind of it), used for decorative purposes, is made by reducing the thin leaves of Dutch metal to a fine powder. Such powders are made of different shades, from dark copper colour to pale gold, by varying the proportion of copper. The grinding is effected with a very little oil, to prevent the metal from being tarnished by oxidation.

Pinchbeck is composed of 3 parts of copper to 1 part of zinc, *Prince's metal*, of equal weights of the two metals. *Mosaic gold* contains about equal parts of copper and zinc. The same name is sometimes applied to a compound of sulphur with tin.

A little tin is added to the brass intended for engraving, since it causes it to break up more easily under the action of the graver. The addition of a little lead (about 3 oz. to 10 lbs.) much facilitates the working of brass at the lathe and with the file, since it prevents the shavings and filings from *greasing* or adhering to the tools.

Brass is liable to be rendered very brittle when placed in situations where it is exposed to continual vibration. This seems to be due to the development of a crystalline structure in the metal, and has occasionally caused the snapping of the suspending chains of chandeliers.

The *lacquering* of brass, in order to protect it from being tarnished by the air, consists simply in varnishing it with shellac dissolved in spirit and coloured with saffron, annatto, dragon's blood, &c., so as to give it a golden hue.

Brass is *bronzed* by coating it with a thin film of arsenic, mercury, or platinum, the last being used only for small articles, such as instruments, on account of its high price. A solution of white arsenic (arsenious acid) in muriatic (hydrochloric) acid, or of corrosive sublimate (chloride of mercury) in vinegar, is brushed over the brass, previously warmed, when the zinc in the brass chemically displaces the arsenic or the mercury from the liquid, and one of these metals is deposited as a coating upon the brass. In bronzing with platinum, a solution of muriate of platina (chloride of platinum) is applied in a similar way. There is much art in obtaining a durable bronze coating of any desired shade of colour. A mixture of chloride of platinum, corrosive sublimate, and vinegar, is used for bronzing the *sights* of guns.

Pins which are made of brass wire are tinned by boiling them with granulated tin, water, and cream of tartar (bitartrate of potash), when the latter, being strongly acid, slowly dissolves the tin, which is afterwards displaced from the solution and deposited upon the brass, because the tin and brass, in contact, form a galvanic couple, which decomposes the salt of tin, precipitating that metal upon the surface of the brass which is the negative plate of the galvanic pair.

In tinning or whitening pins, about 6 lbs. of pins are spread over the bottom of a copper vessel, and covered with 7 or 8 lbs. of grain tin; another layer of pins is then introduced, afterwards more tin, and so on until the vessel is filled. Water is then poured in, the vessel heated, and $\frac{1}{4}$ lb. of cream of tartar sprinkled over the surface. After boiling for an hour the tinning is completed.

Malleable Brass or *Muntz's Metal,* or *Yellow Sheathing.*—This is an alloy of 3 parts of copper and 2 parts of zinc, which differs from common brass in being malleable when hot. It is of course cheaper than ordinary brass, on account of the predominance of the cheaper zinc, and can be more easily rolled into thin sheets. When used for sheathing ships, it keeps cleaner than copper. A small proportion, less than $\frac{1}{100}$th, of lead is now commonly added to Muntz's metal.

The nails employed for securing the sheathing contain, in 100 parts, 87 copper, 4 zinc, and 9 tin, the latter giving them hardness.

Aich Metal, or *Gedge's Metal,* is an alloy of zinc and copper in nearly the same proportions as are contained in Muntz's metal, but it contains also a little iron. It consists, in a hundred parts, of

Copper . .	60·0	Iron . . .	1·8
Zinc . . .	38·2		

This remarkable alloy is very malleable at a red heat, and may be hammered, rolled, or drawn into wire, with the additional advantage of being readily cast. It has been employed in Austria for casting cannon, and some Chinese cannon have been found to consist of a similar alloy.

*Sterro-metal** is another very strong and elastic alloy used by Austrian engineers for the pumps of hydraulic presses. It contains copper, zinc, iron and tin, in the following proportions in a hundred parts, the proportions varying between the assigned limits according to the purpose for which it is required :—

Copper . .	55 to 60	Iron . . .	2 to 4
Zinc . .	34 to 44	Tin . . .	1 to 2

Good specimens of sterro-metal have been found to offer far more resistance than gun-metal to transverse fracture, and it is only two-thirds of the price. It is said that this alloy was accidentally discovered in an attempt to employ, for the manufacture of brass, the alloy of iron and zinc found at the bottom of the zinc-pots in making galvanised iron (p. 167).

A very hard white alloy of 77 parts of zinc, 17 of tin, and 6 of copper is sometimes employed for bearings of the driving wheels of locomotives, and another alloy containing 90 of copper, 5 of zinc, and 5 of antimony, is used for sockets in which the steel or iron pivots of machinery are to work.

German Silver or *Nickel Silver.*—The metal nickel, when alloyed with copper, whitens it more than a similar quantity of zinc. An alloy of copper with a quarter of its weight of nickel has a silvery white colour, and is sometimes called nickel-silver. But the ordinary nickel-silver of commerce contains zinc in addition, and may be regarded as brass whitened by the addition of nickel. The alloy is somewhat troublesome to make, on account of the difficult fusibility of nickel. One method of making German silver consists in melting the copper and nickel together in a crucible, and adding the zinc in pieces previously heated. According to another process, half the copper is placed at the bottom of the crucible, the zinc and nickel are placed

* Named from the Greek adjective *strong, firm.*

upon it, and covered with the remainder of the copper. The crucible is filled up with charcoal-powder, and exposed to a strong heat. The melted metals must be well stirred together with an earthenware stirrer. The proportions of the metals employed in making German silver are varied according to the articles required. For spoons and forks, an alloy of 2 parts copper, 1 nickel, and 1 zinc, is employed. For knife and fork handles, 5 copper, 2 nickel, and 2 zinc. For rolling into sheets, 3 copper, 1 nickel, 1 zinc. For castings of German silver, 3 parts of lead are added to 100 parts of the alloy first mentioned. The larger the proportion of zinc, the more liable is the German silver to assume, after a time, the yellow colour which is so objectionable.

German silver was originally made from an ore containing copper, nickel, and zinc, found at Suhl in Germany.

In Nassau an alloy of copper and nickel is extracted from a complex ore consisting of copper pyrites, iron pyrites, spathic iron ore, hæmatite, and sulphuret (sulphide) of nickel. The treatment of the ore is similar in principle to the Welsh method of smelting copper-ores, except that the regulus, corresponding to fine metal (p. 117), which contains both the copper and nickel, is completely roasted until all the sulphur is expelled, and both metals are converted into oxides which are afterwards reduced to the metallic state by charcoal, yielding an alloy of copper with about half its weight of nickel, which is well suited for the manufacture of German silver.

At Mansfeld, advantage is taken of the chemical attraction which exists between nickel and arsenic, to separate the nickel from its ore in the form of *speiss*, which is a semi-metallic *matt* composed of nickel and arsenic. The greater part of the latter is separated by roasting, which also converts the nickel into an oxide; this is afterwards treated by a purely chemical process to separate the rest of the arsenic, and the oxide of nickel is eventually reduced to the metallic state by heating it with carbonaceous substances.

An alloy of 7 parts of copper with 1 part of nickel has been employed for small coinage in America.

Aluminium-bronze is an alloy of nine parts of copper and one part of aluminium. In colour, it much resembles gold, but is much harder and lighter. It is extensively used as a cheap imitation of gold, but it becomes tarnished in course of time. It has also been employed instead of steel for perforating postage-stamps, etc., and is said not to be so soon blunted.

LEAD.

Whether lead in the metallic state has ever been found as a true natural product appears to be doubtful, since the small quantities which have been found associated with the ores of lead may have been accidentally reduced.

Although minerals containing lead are pretty abundant, there are only two which are found in sufficient quantity to serve as sources from which to extract the metal on the large scale.

Ores of Lead.

	Composition	Lead in 100 parts of the pure ore
Galena or Sulphuret of Lead .	Lead, Sulphur	86½
White Lead Ore or Carbonate of Lead .	Lead, Oxygen, Carbonic Acid	77½

Galena is by far the most abundant of the compounds of lead. It forms extensive veins, traversing clay slate in Cornwall, and limestone in Derbyshire and Cumberland. It is also found in Flintshire (Holywell), Scotland (Leadhills), and the Isle of Man. Spain yields abundance of galena in Catalonia, Grenada, and at Linares in the Sierra Morena, where it occurs in granite. This ore is also abundant in the Upper Hartz, at Freiberg in Saxony, and in the United

States of America. Few ores are so easily recognised at once as galena ; it is distinguished by its lustre, which is almost metallic, its dark grey colour, and its great weight (specific gravity, 7·5). It can generally be easily split up into rectangular fragments, and often occurs in distinct cubical crystals of large size.

Galena almost invariably contains silver, which takes the place of a part of the lead in its combination with sulphur, without producing any alteration in the crystalline form and general appearance of the ore. A galena containing two parts of silver in a thousand would be spoken of as an *argentiferous galena,** because even that small proportion of metal can be profitably extracted from the lead after smelting it from the ore.

Antimony is also found in many specimens of galena, as a sulphuret of antimony, and its presence has a serious influence upon the quality of the lead extracted from the ore. The minerals commonly associated with galena in the vein are blende (sulphuret of zinc) and copper pyrites, whilst *cawk* or heavy spar (sulphate of barytes), calc-spar (carbonate of lime), and fluor-spar (fluoride of calcium) are often found adjacent.

White-lead ore or *carbonate of lead* is a much less important ore, often occurring in veins of galena, and apparently produced by a chemical alteration of this ore. When pure, it is a white crystalline mineral, but it has often an earthy appearance, and it is so unlike galena that miners have been known to reject it as worthless. Sometimes it has a dark colour, from the presence of a little galena intermixed with it. Carbonate of lead is found in considerable quantity near Aix-la-Chapelle, as well as in Spain, and in the valley of the Mississippi. This ore is so seldom smelted apart from galena, that it is not necessary to describe its treatment separately.

* *Argentum,* Latin for silver ; *fero,* I bear.

Sulphate of lead (composed of lead, sulphur, and oxygen), or *Anglesite*, is very rarely found in any quantity. Australia furnishes some of it, containing a considerable proportion of silver.

In order to prepare the lead-ore for smelting, it is sorted by hand, the worthless pieces being rejected, and broken up, either with a hammer or between crushing-cylinders; it is then washed, in much the same way as the ore of tin (p. 131), in order to separate, as far as possible, the foreign matters mingled with it. The differences in the ore have led to the adoption of different methods of conducting the operation of smelting; thus in Derbyshire and Flintshire, where the lead ores are rich and contain very little quartz (silica), the galena is smelted in reverberatory furnaces, whilst at Alston Moor, and generally in the lead-works of the North, small blast furnaces are employed.

Smelting of Galena in the Reverberatory Furnace.—The chemical principles upon which metallic lead is separated from galena are similar to those involved in the last stage of the extraction of copper (roasting the fine metal for blistered copper, p. 118), the sulphur being finally expelled in the form of sulphurous acid, produced by its combination with oxygen previously taken up from the air. The galena is first roasted until a part of it has become converted into oxide of lead, its sulphur having combined with oxygen and been removed as sulphurous acid.

Galena (*Lead* *Sulphur*)	and	*Oxygen* from the air	give
Oxide of Lead (*Lead* *Oxygen*)	and	*Sulphurous Acid Gas* (*Sulphur* *Oxygen*)	

During this roasting process, another portion of the galena is converted by the oxygen of the air into sulphate of lead.

When the roasting has proceeded far enough, the oxide of lead and sulphate of lead are melted with that portion of the galena which has escaped alteration, when the whole of the sulphur is converted into sulphurous acid, and the lead is left in the metallic state.

Galena (*Lead* *Sulphur*) and *Oxide of Lead* (*Lead* *Oxygen*) give

Sulphurous Acid Gas (*Sulphur* *Oxygen*) and *Lead*

Again—

Galena (*Lead* *Sulphur*) and *Sulphate of Lead* (*Lead* *Sulphur* *Oxygen*) give

Sulphurous Acid Gas (*Sulphur* *Oxygen*) and *Lead*

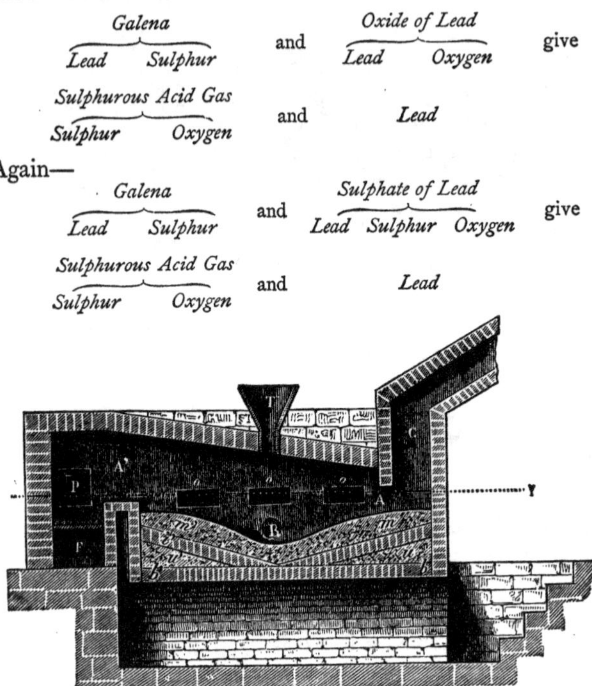

FIG. 62.—Reverberatory Furnace for smelting Galena.

The reverberatory furnace in which the smelting of galena is effected is represented in Figs. 62, 63. The hearth (B) is about eight feet by six, and is separated from the grate (F) by a fire-bridge which rises to within about eighteen inches from the arch (AA′), the latter gradually descending, as it approaches the chimney, until it is within about six inches of the hearth. The flame and products of combustion, after passing over the hearth, are conducted by two openings into a flue about eighteen inches wide; this flue makes a bend downwards towards the top, and is carried into a chimney between fifty and sixty feet high. The flue is so constructed that it may be readily opened to clear out the deposit from the lead fumes. The fire-door (P), for throwing the coal upon the grate, and the ash-pit (F) are on opposite sides of the furnace; that upon which the fire-door is situate is called the *labourer's side*, whilst that opposite is the *working side.*

On the labourer's side, there are three openings (*o*), about six inches square, at equal distances, which can be closed when necessary with iron plates. There are three corre-

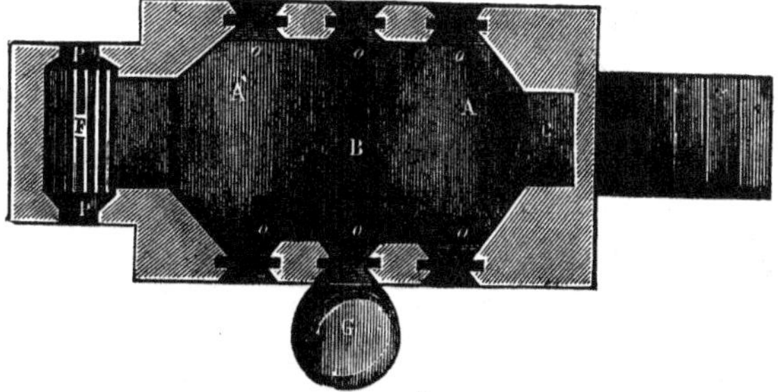

FIG. 63.—Plan of Reverberatory Furnace for smelting Galena.

sponding openings on the working side of the furnace, as well as two tapping-holes for the lead and slag respectively.

The hearth of the furnace is lined with the slags from previous operations, which are spread over it while in a pasty state, before solidifying, and fashioned to the proper shape as shown in Fig. 64. On the labourer's side, it is nearly up to the level of the working doors, but on the opposite side it is hollowed out so as to be eighteen inches below the middle door; this being the lowest part of the hearth, where the melted lead collects, a tap-hole is provided for running off the metal, and at some distance above it is the aperture for the escape of the slag. Adjacent to the tap-hole there is a basin outside the furnace for the reception of the lead. The ores are selected, if possible, so that the earthy matters associated with them may act as fluxes to each other (p. 35), or else some lime is added in order to combine with the silica and form a fusible slag.

The operation of smelting galena in the reverberatory furnace consists of four consecutive stages, distinguished as first, second, third, and fourth *fires*.

First fire.—As soon as the lead smelted in the preceding

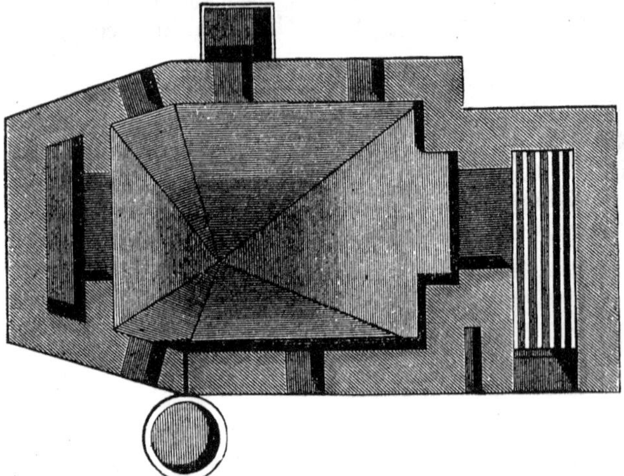

FIG. 64.—Hearth of Reverberatory Furnace for smelting Galena.

operation has been tapped into the outer basin, and while the furnace is still glowing, the fresh charge of about a ton of ore is introduced through a hopper (T, Fig. 62) in the arch of the furnace. No regular fire is made up, but only a little coal is thrown into the grate to keep up a moderate temperature, for if this were raised too high at first, the galena would fuse, and the roasting would be rendered impossible. A workman stationed at the labourer's side spreads the ore uniformly over the surface with a rake, after which the doors are closed and the draught moderated by lowering the damper. Since there is little fuel upon the grate, a considerable quantity of unconsumed oxygen of the air passes over the hearth of the furnace, so that some oxide of lead and sulphate of lead are soon formed.

After the roasting has been continued for some time, the skimmings from the lead, run into the outer basin at the end of the last smelting, are thrown into the hearth. These skimmings consist chiefly of a combination of lead with a small quantity of sulphur (subsulphuret or disulphuret of lead), which is speedily acted on by the oxide and sulphate of lead, as above explained, with separation of metallic lead which runs down into the hollow, and is drawn out through the tap-hole into the basin. This first portion of lead contains a larger proportion of silver than that tapped at a later period of the process. The workman occasionally turns over the ore to expose fresh surfaces, and, if necessary, throws a little small coal upon the charge to prevent the oxidation from being carried too far. About an hour after the commencement, a large quantity of lead is run off, being chiefly derived from the action of the skimmings upon the roasted galena.

After an hour and a half from the commencement, all the doors are thrown open, and the ore is well turned over by two workmen placed on opposite sides of the furnace, after which the doors are closed. At the end of two hours, the first fire is completed, a sufficient proportion of the galena having been converted by the roasting into oxide and sulphate of lead.

Second fire.—The damper of the furnace is now partly raised, and more coal is thrown into the grate, so as to bring the temperature up to a bright red heat.

The sulphuret of lead in the ore now acts upon the oxide and sulphate formed during the previous roasting, and the melted lead begins to run out in abundance. The workman stationed on the working side thrusts the pasty slags out of the basin, whilst the man on the labourer's side spreads them over the rest of the hearth; a little quicklime is now thrown in to assist the fluxing, and the doors are left open for a short time to cool the hearth a little, which is found to facilitate the draining out of the lead, probably by chilling and partly solidifying the slag. Occasional cooling is desirable also for converting the mass into a pasty condition, for if it were allowed to become liquid, the unaltered sulphuret of lead would sink to the bottom instead of being acted on by the oxide and sulphate of lead. The time occupied by this second fire is about an hour.

Third fire.—The doors are all shut, and the damper entirely opened, more fuel being thrown upon the grate so as to raise the hearth to a still higher temperature for about three quarters of an hour, when the doors are again opened, the slags spread over the hearth, and a fresh quantity of lime thrown upon them. The lime enters into combination with any silica which may have united with the oxide of lead, and sets the latter free to act upon any portions of unaltered sulphuret of lead. The lime also acts advantageously by diminishing the fusibility of the mass and thus facilitating the contact between the sulphuret of lead and the oxide. This third fire also occupies about an hour.

Fourth fire.—The grate is again charged with fuel and the doors closed for about three quarters of an hour, the furnace being thus raised to its highest temperature. The tap-hole is then opened to allow the lead to run into the outer basin (G, Fig. 63), and some lime is mixed with the slags in order to *dry up* or partly solidify them, when they are raked out through the openings on the labourer's side, and the furnace is ready to receive a fresh charge of ore. A little small coal is sometimes thrown upon the hearth at the conclusion of the fourth fire, to remove the oxygen from any oxide of lead which may still remain.

The iron of the tools employed in stirring the contents of the hearth is seriously corroded by the sulphur in the ore.

The whole operation of smelting in the reverberatory furnace lasts about five hours, and the coal consumed is about 12 cwts. for every ton of ore, being less when fluor spar is present to act as a flux.

The lead in the outer basin is covered, as already mentioned, with a layer of subsulphuret of lead, which is thrown in during the first fire with the next charge. Logs of wet wood are sometimes let down into it, as in the case of tin, to cause ebullition and promote the separation of the slags, before the lead is cast into pigs.

The slag amounts to about one-fourth of the weight of the ore, and sometimes contains as much as 40 parts of lead in the hundred, so that it is smelted in the *slag-hearth*, to be described hereafter.

At Bleiberg in Carinthia, the lead is extracted from galena by a process much resembling that just described; but wood is employed as fuel, the grate being at the side instead of at the end of the hearth, and the hearth of the furnace is a single inclined plane, allowing the reduced lead to flow at once out of the furnace. The first portion of lead which runs out is known as *virgin lead*, and is purer than the *pressed lead* obtained later in the process when the temperature is much higher. The Carinthian lead (*Villacher* lead, from the town of Villach) is in high repute for its purity.

In Nassau, a furnace more nearly resembling the English reverberatory is employed, and towards the end of the process, some green wood is added to the charge upon the hearth, in order that the steam and gases evolved from it may agitate and mix the pasty mass. The lead collected in the basin outside the furnace is stirred with wood (like tin, see p. 139) before being run into pigs.

In some of the Continental furnaces, metallic iron is added to the charge in order to combine with the sulphur in the galena, and separate the lead in the metallic state.

When galena containing much antimony is smelted in the reverberatory furnace, a portion of the oxide of lead combines with the oxide of antimony to form a compound which can only be decomposed by the coal at a very high temperature, so that the first portions of lead obtained are much purer from antimony than those at the end of the process.

Smelting of Lead Ore in the Scotch Furnace or *Ore-hearth.*—Since this is a blast furnace, it is found advantageous to roast the ore before smelting it, in order that it may be rendered more porous and may offer less obstruction to the blast.

The ore is spread over the hearth of a reverberatory furnace, not unlike that employed for roasting copper ores (p. 109), in charges of about half a ton, and roasted at a moderate heat for about eight hours, being frequently turned over as is usual in roasting operations. Some antimony is thus expelled from the ore, which would otherwise harden the lead; a considerable quantity of sulphur also burns off. The roasted ore is raked out of the furnace into a pit filled with water, which causes it to fly into fragments suitable for charging into the ore-hearth.

The ore-hearth (Fig. 65) is a small square forge or blast furnace about two feet high, and 18 inches by 12 internal area. It is arched over at the top, so that the *lead fume* may be conducted into a long flue, sometimes five feet high and three feet wide, in which a large quantity of oxide of lead and sulphate of lead is deposited. At the Allenheads works, this flue is carried up the side of a hill for three miles before it terminates in the chimney, in order to secure perfect deposition of the lead fume, which would otherwise involve very considerable waste; for although lead is not, like zinc, a metal capable of being distilled, both the metal and its sulphuret, when heated in a strong current of air, are liable to be carried off in the form of vapours, which afterwards combine with oxygen from the air, and are deposited as oxide and sulphate in the flues; these deposits are afterwards heated in the calcining furnace till they can be made to stick together, and are then smelted in another furnace called the *slag-hearth.* A *rain chamber* is often provided, in which the condensation of the fume is assisted by water ; and the great length of flue renders it necessary to assist the draught by large exhausting pumps.

Since a very moderate temperature is required in this furnace, the sides and bottom are lined with cast-iron plates, and in front of the furnace, where there is an opening about

Fig. 65.—Scotch Furnace or Ore-hearth for the extraction of Lead.

a foot high, a sloping iron plate *a b* (*work-stone*) is fixed, upon which the materials can be raked out when necessary, for examination and manipulation by the smelter. The bottom of the furnace, upon which the melted lead collects, is about 4½ inches below the upper surface of this iron plate, in the edge of which there is a groove cut, near to the side of the furnace, through which the melted lead may run when it rises to a sufficient height, into a gutter (*f*) which conveys it into a cast-iron pot (F) heated by a separate fire, and called the *melting-pot.*

The blast-pipe enters at the back of the furnace, about 11 inches from the bottom. Peat is the principal fuel employed in the furnace, in the form of square blocks, with which the furnace is filled at the commencement, some judgment being required in their arrangement, and the fire is lighted by placing one of the blocks, already kindled, in front of the blast-pipe, when the combustion soon spreads throughout the furnace. The first charge introduced into the ore-hearth does not consist of the roasted ore, but of the residue from a previous smelting operation, which is called *browse,* and consists of partly reduced ore mixed up with cinders; before this is thrown in, a little coal is put on the fire to raise the temperature, and in a short time the charge is raked out upon the work-stone in front of the furnace, and examined, in order that the *grey slag,* a shining glassy mass, may be picked out and thrown on one side. This grey slag contains a quantity of silicate of lead, with silicate of lime, &c., and requires a higher temperature for the extraction of its lead than is attainable in this furnace; it is therefore smelted in the *slag-hearth,* to be presently noticed.

The browse cleaned from slag is thrown back into the furnace, and its behaviour observed; should it appear to melt too readily, it is rendered less fusible by adding a little lime, lest it should run down to the bottom of the furnace with the metallic lead; on the other hand, if it does not become soft enough to permit the lead to separate, lime must also be added to soften it. These apparently opposite effects of the addition of lime will be intelligible on referring to the remarks upon the use of that material in the extraction of cast iron (p. 35). A fresh quantity of grey slag is thus formed, and is removed by the workman.

A peat is now placed before the opening of the blast-pipe, in order to prevent any dust from entering it, and a quantity of roasted ore with a little coal is thrown in. After about twenty minutes, the charge is again raked out on to the work-stone, the grey slag picked out, the remainder thrown back into the furnace, and a fresh charge of roasted ore and coal added.

These operations are repeated during 14 or 15 hours, in which period one or two tons of lead will have collected in the outer basin, according to the richness of the ore, and the proportion (varying from $\frac{1}{10}$th to $\frac{1}{15}$th of the whole) which has been removed in the grey slag.

The separation of the metallic lead is partly due to the action of the sulphuret upon the sulphate and oxide of lead, as explained at p. 178, and partly to the removal of oxygen from the oxide by the carbon of the fuel.

The lead extracted in the ore-hearth is purer and softer than that obtained by the reverberatory furnace, the temperature being so low that the other metals contained in the ore are not reduced.

Smelting of Slags, &c., in the Slag-hearth.—In this operation the object is to extract as much of the lead as possible from the slags and other residues, without reference to its purity, by the employment of a very high temperature, so as to completely liquefy the slag. The general construction of the furnace (Figs. 66, 67) is not very different from that of the ore-hearth, but it is larger, being 3 feet high, and 26 inches by 22, internal area. The sides are built of sandstone, in order to resist the much higher temperature of this furnace. The bottom (A) of the furnace consists of a cast-iron plate, and is covered with a layer, about 16 inches thick, of porous cinders tightly rammed down, which serves as a strainer to separate the lead from the slag, since the melted metal easily percolates into the porous cinders, which protect it from being oxydised again by the air, and runs thence into a receptacle (B) outside the furnace, which is also filled up with similar cinders, and has an opening through which the lead flows into an iron pot (E) kept hot over a separate fire. The slag runs off the surface of the layer of cinders, both in the furnace and in the receiving basin, and falls into a cistern of water (C), where it breaks up, so that the lead entangled in it is easily separated by washing. The fire is lighted with peats as in the ore-hearth, the

FIG. 66.—Slag-hearth for extracting Lead.

blast being forced through a nozzle at the back of the furnace, about four inches above the layer of cinders. Some coke is then thrown in, and about six hours after, when the temperature is sufficiently high, a charge of the slags, &c., which are to be smelted. Coke and slags are thus added in

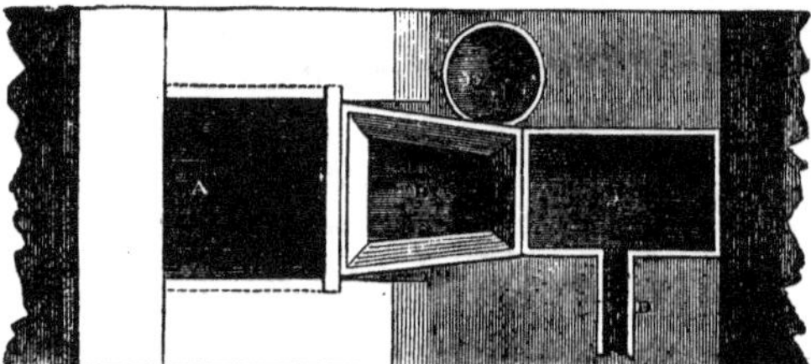

FIG. 67.—Plan of Slag-hearth.

alternate charges, as in the iron blast-furnace, until the furnace requires repair.

The charge for the slag-hearth commonly consists of:

100 parts of slag from the reverberatory furnace;
20 ,, coal-ashes;
13 ,, clay-hearths of old furnaces, impregnated with lead;
5 ,, rich slag from a previous operation.

The silica and alumina present in the clay and in the coal-ashes combine with the lime and oxide of iron in the slag from the reverberatory furnaces, and form an easily-fusible slag. The lead is reduced to the metallic state mainly by the action of the heated carbon, which removes the oxygen from the oxide of lead. The coke is piled up towards the front, and the charge towards the back of the furnace, and a *nose* or prolongation of the tuyère is allowed to be formed by the solidified slag, so as to carry the blast up the centre of the furnace. When a cold-blast is employed, this nose is apt to become too long, so that air heated to about 300° F. is found to answer better, beside effecting a considerable saving of fuel. If the blast is too hot, the slag will not be chilled so as to form a nose.

A very inferior description of lead (*slag-lead*) is obtained from the slag-hearth.

In some parts of Spain, lead is extracted from the slags of the Roman lead-furnaces.

Richardson's Furnace, or the *Economico Furnace*, which is employed in Newcastle and the neighbourhood instead of the slag-hearth, as well as for the extraction of lead from the ore, is a modification of the *Castilian furnace* (Fig. 68), being also a blast-furnace, in which the blast is either supplied by a blowing-engine through three tuyères or blast-pipes, or is drawn into the furnace through five or six openings, by the action of a small chimney. The body (A) of the furnace is circular and is built of fire-brick, about 8½ feet high and 2½ feet in diameter, the bottom (B) being lined with a mixture of clay and powdered coke, well beaten down, and hollowed out to receive the lead. The ore, or mixture of ore and slag, smelted in this furnace, must not contain more than 30 parts of lead in the hundred. The ore is roasted previously to its introduction into the furnace, which is charged with ore and fuel through an opening (I) in the square brick structure supported on four pillars, which surmounts the furnace, and, if necessary, the charge is sprinkled with water from a rose, to

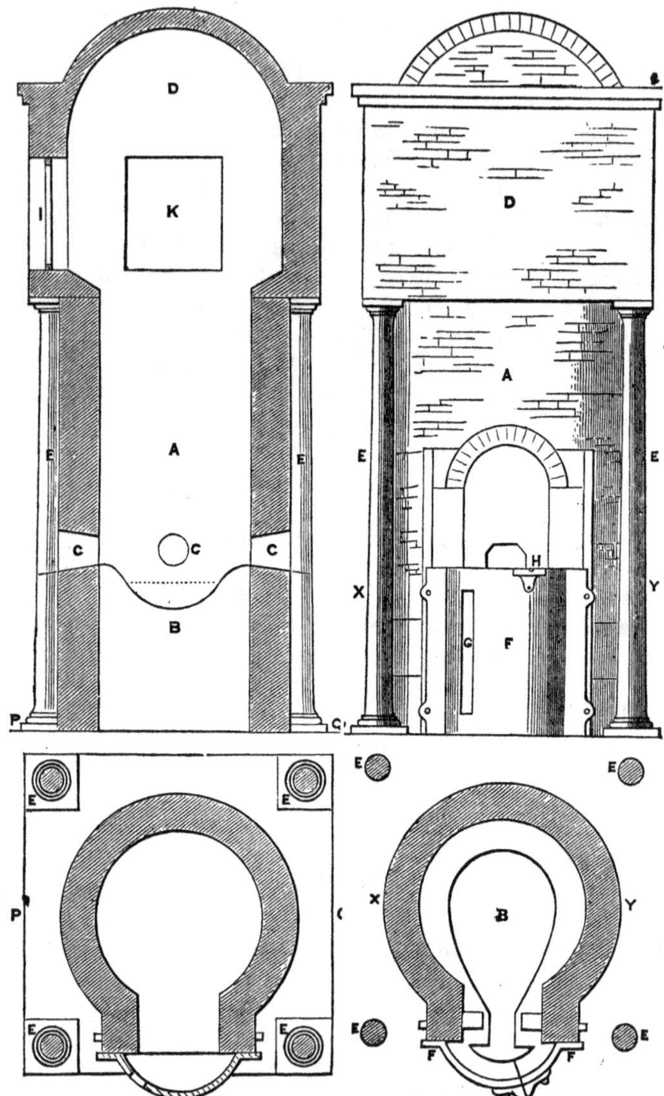

FIG. 68.—Castilian Furnace for Lead-smelting. A, Body of the furnace. B, Crucible for receiving the lead. C, Blast-pipe. D, Masonry enclosing the top of the furnace. E, Cast-iron pillars. F, Receptacle for the lead. G, Slot for the tapping-hole. H, Spout for overflow of slag. I, Charging-door. K, Flue. P Q, Ground line.

prevent the dust from being carried into the flue. The slag flows over the side of the hearth, as in an iron blast-furnace, into cast-iron waggons, whilst the lead accumulates in the cavity at the bottom, and is tapped out from time to time into an iron basin (F). Limestone is sometimes mixed with the charge, to flux the siliceous matters.

At Clausthal in the Hartz, the galena is reduced by fusing it in a small blast-furnace or *cupola-furnace* with granulated cast-iron, which combines with the sulphur to form a sulphuret of iron, and sets the lead at liberty. The sulphuret of copper which is present in the ore is not decomposed in the process, but forms a *matt* upon the surface of the lead, and after converting the sulphuret of iron into oxide by roasting, and removing the oxide by fusion with siliceous matters, the sulphuret of copper is sent to the copper smelting-works. At some works, slags from the refining of iron (p. 52) are employed to assist in the decomposition of the sulphuret of lead. When ores in a finely divided state form part of the charge of the slag-hearths and cupola furnaces on the Continent, they are often mixed with clay or lime and moulded into bricks before being thrown into the furnace.

Softening of Lead in the Calcining or Improving Furnace.—The lead obtained by either of the above processes sometimes contains considerable quantities of silver, antimony, copper, and iron, which harden the metal and render it unsuitable for some of its applications. English lead is the purest which is to be found in commerce, and Spanish lead is the most impure ; the composition of two samples in 100 parts is here contrasted :

	English	Spanish
Lead	99·27	95·81
Antimony . . .	0·57	3·66
Copper	0·12	0·32
Iron	0·04	0·21

Even the English specimen is sufficiently impure to be designated a *hard lead.*

When the lead contains any notable proportion of silver, it is treated by a special process to be described hereafter, but when the hardness is due to the presence of antimony, &c., the lead is softened or improved by exposing the melted metal in a very shallow pan to the action of the oxygen of the air, which converts the antimony, copper, iron, and a considerable portion of the lead, into oxides, which collect as a dross upon the surface, and are skimmed off at intervals, until the lead is found to be sufficiently softened.

The improving furnace (Figs. 69, 70) is a reverberatory furnace with a low arch, 18 inches above the hearth near the fire-bridge, and 6 inches near the chimney, towards which the flame is drawn by two flues (F). Since a large

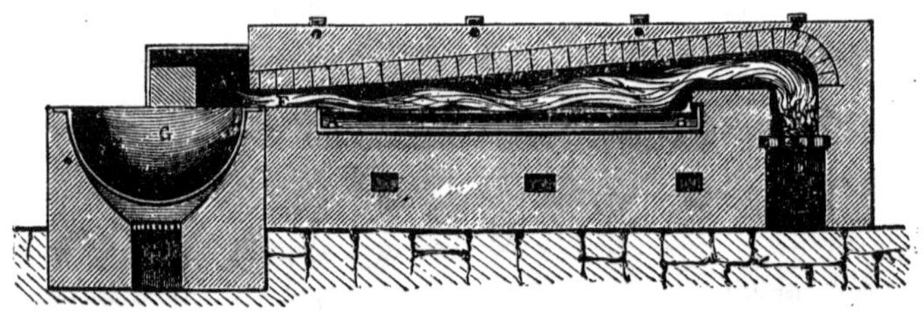

FIG. 69.—Calcining Furnace for improving Hard Lead.

spreading flame is required, the fireplace (D) is 5 feet long and 18 inches wide, being divided from the hearth by a fire-bridge 27 inches wide and 16 inches above the hearth (B). In the hearth of the furnace there is set, with a space round it to allow for expansion, a cast-iron pan measuring 10 feet by 5, which is 8 inches deep at the end nearest the grate, and 9 inches at the other end; at this deepest end there is an iron gutter (*f*) stopped up during the process, by an iron plug with a weighted lever through which the lead may be run out when it is sufficiently refined.

Eight or ten tons of the hard lead are melted in an iron pot (G) and ladled into a gutter (H) through which they run into the improving pan, which has been already heated to dull redness; the gutter is then closed by a damper, and the improving process commences. Its duration depends upon the amount of the foreign metals (especially of antimony) which the lead contains, a single day's calcination being sufficient to soften some leads, whilst others require two or three weeks' exposure to the action of the air which passes through the furnace. The dross is raked off occasionally, so as to expose the surface of the metal, and the progress of the refining is observed by ladling a small sample into an ingot-mould, when its surface assumes a peculiar crystalline appearance if the refining is completed. The

FIG. 70.—Plan of Improving Furnace.

refiner also judges that the lead is sufficiently softened, if a rainbow or iridescent film is formed upon the surface when a rake is pushed over it from the working door. The softened lead should not form round globules when poured upon a heated iron plate.

The dross, which consists chiefly of the oxides of antimony and lead, is mixed with coal and ground under edge-runners, previously to being smelted in a small reverberatory furnace, the hearth of which gradually slopes down towards the chimney, where there is a cavity for the reception of the metal, which constantly flows out through the tap-hole into an iron pot, to be afterwards transferred to the pig-moulds.

The hard lead thus obtained is either again calcined with a fresh portion of metal, or if it contains a very large proportion of antimony (of which some specimens contain a third of their weight), it is sold to the type-founders.

The process of improving the moderately hard lead produced at Altenau in the Upper Hartz resembles the *boiling* of tin, and consists in melting about 11 tons of the metal in an iron pot $5\frac{1}{2}$ feet deep and 3 feet wide, and stirring it for two hours with a birchen pole moved by machinery, when the violent bubbling of the gases through the metal continually renews the surface in contact with the air, causing the formation of dross containing the impurities.

In France, the improving furnaces often have two fires, one at each end of the pan.

Another method of improving very hard lead consists in melting it, as above, in a cast-iron pan, and throwing upon the skimmed surface a small quantity of a mixture of Peruvian saltpetre (nitrate of soda), soda and lime, the addition being repeated until the metal is sufficiently softened; the oxygen of the saltpetre converts the antimony into antimonic acid which combines with the soda and lime, and is removed as dross, together with a considerable quantity of oxide of lead.

Refining of Lead containing Silver by Pattinson's Process.—This very simple and beautiful process, which was introduced in 1829, has not only greatly improved the quality of lead, but has very much increased the production of silver in England; previously to that date no process existed by which the silver could be removed so as to leave the lead in the metallic state, and it was necessary to convert the whole of the lead into an oxide in order to separate the silver, this oxide being afterwards smelted to recover the lead. Since this could not be made to pay unless the lead contained at least eleven ounces of silver in the ton, any smaller quantity of silver was left in the lead sent into the market, the lead being thereby hardened, and the silver entirely lost for all useful purposes. After the introduction of Pattinson's process, much of the old lead was eagerly bought up for the sake of the silver which it contained.

The process depends upon the property of lead to crystallise at a lower temperature than an alloy of lead and silver, so that if melted lead containing a small proportion of silver be allowed to cool slowly and constantly stirred, the small crystals of lead which are formed at first will contain little or no silver, that metal remaining in the liquid portion.

To carry out this principle, a series of melting-pots is employed. The number of pots as well as their form and general arrangement will differ somewhat in different esta-

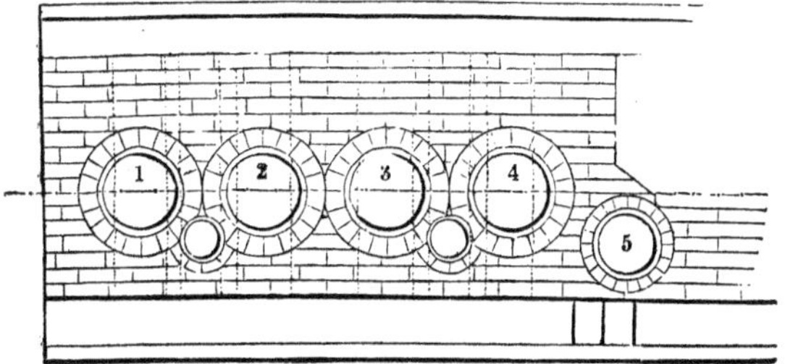

FIG. 71.—Pots for desilverising Lead.

blishments, but for the sake of illustration, a *desilverising* plant containing five pots may be taken (Fig. 71). These pots are made of cast iron and set in masonry; the *working-pots*, 1, 2, 3, and 4, are oval in shape, their mouths being 40 inches by 26, and they are shaped at the bottom like the small end of an egg. 5 is the *market-pot* for melting the desilverised lead before casting it into pigs; it is smaller than the others. The smallest pots (about two feet in diameter), between 1 and 2, and between 3 and 4, are the *temper-pots*, for containing the melted lead in which the perforated ladle (Fig. 72) is warmed, which is used for fishing out the crystals of lead. This is an iron ladle about 18 inches wide and 5 inches deep, with an iron handle of 4½ feet and a wooden handle of about 5 feet in length; the holes in the ladle are $\frac{1}{2}$ inch wide and $\frac{3}{4}$ inch apart. Each pot is heated by a separate fire.

The lead to be refined is usually in pigs (or *salmons*) weighing from 120 to 140 lbs. each. About 64 of these (or 4 tons) are melted in pot 1. (See Fig. 73.) When they are perfectly melted, the fire is raked out, and the oxide is skimmed from the surface of the lead. In order to hasten the cooling of the metal, one or two pigs of cold lead are thrown in, or a little water is thrown upon the surface so as to form a solid crust, which is then pushed down into the liquid metal. This is continued until crystals of lead begin to form.

Fig. 72.—Ladle for fishing out Lead-crystals.

The workman then detaches any lead which has solidified on the sides of the pot, and stirs the melted metal with an iron bar in order to preserve an equal temperature throughout. Another workman takes the perforated ladle out of the temper-pot in which it has been heated, and fishes up the crystals which have formed. The handle of the ladle is then rested upon a pig of lead faced with iron placed at the edge of the pot to serve as a fulcrum, and the workman seizes the end of the long handle, and jumps down from the platform around the pots on to the floor, thus tilting the ladle up out of the melted lead, over which he shakes it violently so as to drain all the liquid metal back into the pot. The ladle is then swung by a crane over pot 2, into which the crystals are thrown; after this has been repeated for an hour, only about one ton of lead richer in silver is left in pot 1, the quantity being ascertained by trying the depth of the metal in the pot.

The removal of the crystals is still proceeded with, but

since these will now contain too much silver to be introduced into pot 2, they are thrown upon the ground in order to be afterwards melted up with more lead in pot 1. When only $\frac{1}{2}$ ton of the rich liquid alloy is left in pot 1, it contains about three times as much silver as the original lead, and is ladled out and cast into eight pigs, which often contain as much as 150 ozs. of silver in the ton, together with any copper and

FIG. 73.—Desilverising Lead by Pattinson's Process.

antimony which were contained in the original lead; whilst any arsenic which was present will have passed into the crystals. The half-ton of crystals which have been thrown upon the ground are now melted in pot 1 with a fresh quantity of the original lead, and treated as before.

The three tons of crystals of lead poor in silver which were transferred to pot 2, are made up to four tons by adding lead of the same richness in silver, and submitted to a repetition of the same treatment, about three-fourths of it being transferred, in crystals, to pot 3, one-eighth of it, in the form of richer crystals, being thrown upon the ground to be remelted in pot 2, and the remainder, which is left in the liquid state at the bottom, is ladled out into pot 1.

The crystals in pot 3 are treated in the same way, the portion remaining liquid being transferred to pot 2, and the poorer crystals melted in pot 4. Finally the crystals of poor lead formed in pot 4 are ladled into pot 5, to be cast into pigs, which are treated again, if necessary, until the silver is at last reduced to half an ounce in the ton. By a single operation in the four pots, as just described, the silver in the marketable lead is reduced to one-tenth of its original amount.

The process of cupellation for extracting the silver from the rich lead is described under Silver.

The quality of the lead is greatly improved by Pattinson's process, not only because the bulk of the antimony and copper remain with the silver in the liquid portion, but because these and other impurities which tend to harden the lead are converted into dross by the oxygen of the air, precisely as in the improving or calcining process.

The description just given refers to an operation with Pattinson's process, upon the *low system*, as it is termed, in which $\frac{7}{8}$ths of the contents of each pan are removed in crystals, whilst according to the *high system* only $\frac{2}{3}$rds are removed.

The high system is preferred for the treatment of the leads richer in silver and of otherwise impure quality, as many as fifteen pans being sometimes employed, so that there is a greater chance of oxidising the impurities. Both systems are combined in some establishments, in order to suit the different descriptions of lead. In working the high system, it is generally found that the crystals taken out of a given pot contain about half as much silver as the alloy originally contained; thus nine tons of a silver-lead containing ten ounces of silver to the ton would yield six tons of crystals containing only five ounces to the ton, and three tons of liquid alloy containing twenty ounces to the ton.

Where many pans are employed, they are commonly hemispherical, being about 5 feet wide and 2½ feet deep.

If the lead contains as much as 60 ozs. of silver in the ton, the third pan is selected for melting it, but if it contains only seven ounces in the ton, it is melted in the 7th pan, an intermediate pan being selected in other cases.

As much as twelve tons of lead are sometimes melted, to begin with, upon the high system, and two-thirds of it are ladled out in crystals into the next left-hand pan—an operation requiring about two hours. The liquid alloy is then chilled into a pasty condition by throwing water upon it, and ladled out into the next right-hand pan, the melting-pot being again charged with twelve tons of the original lead; after a sufficient quantity of metal has accumulated in the adjacent pots, the crystallising operation is repeated with their contents, the two-thirds of crystals being always transferred to the next left-hand pan, and the one-third of liquid alloy to the next right-hand pan. Supposing, therefore, the rule to hold good, that the crystals contain half as much, and the liquid twice as much silver as the silver-lead from which they were obtained, the lead would be found, after passing through four pots to the right of the melting-pot, to contain sixteen times as much silver as the original lead, whilst, after it had passed through four pots to the left of the melting-pot, it would contain only one-sixteenth of the original proportion of silver, as indicated below, where a lead containing seven ounces of silver in the ton is supposed to have been melted originally in pot No. 7.

No. 1.	No. 2.	No. 3.	No. 4.	No. 5.	No. 6.
448 oz. per ton.	224 oz. per ton.	112 oz. per ton.	56 oz. per ton.	28 oz. per ton.	14 oz. per ton.

No. 7.	No. 8.	No. 9.	No. 10.	No. 11.
7 oz. per ton.	3½ oz. per ton.	1¾ oz. per ton.	⅞ oz. per ton.	$\frac{7}{16}$ oz. per ton.

Instead of waiting until the proper quantity of metal for crystallising has accumulated in a given pot, the right complement of a sample of lead of the richness suitable to that pot is commonly added from the stock kept in the establishment.

In some works, Spanish lead is desilverised by mixing it, in a liquid state, with melted zinc, which is afterwards allowed to rise to the surface, carrying with it nearly the whole of the silver; the latter may be recovered either by distilling the zinc away, or by dissolving it in sulphuric or hydrochloric acid.

Uses of Lead.—Many of the uses of lead result from its softness and plasticity, properties which it possesses in a higher degree than any other metal commonly used in the metallic state. The ease with which it may be rolled into sheets recommends it for roofing and for lining sinks, cisterns, &c., particularly since it can be easily adapted to any shape with the aid of a mallet. Again, its softness enables it to be made into pipes either by casting a very thick cylinder round an iron core, and drawing it through progressively diminishing steel dies, or by forcing the melted metal, by hydraulic pressure, through a steel cylinder with a core, from which the solidified metal issues with the required form and dimensions.

The great weight of the metal (specific gravity 11·4) is unfavourable to its employment for roofing, and its ready fusibility (at 620° F.) is another disadvantage, the terrors of a conflagration being sometimes aggravated by the pouring down of the melted lead from the roof.

The poisonous nature of the compounds of lead renders it dangerous to use it for cisterns and pipes with which water, and especially soft water, is to remain in contact for any considerable period, for although lead itself is not acted upon by water, the oxygen of the air, which is always held in solution by water, readily converts a portion of the lead into an oxide of lead, which is dissolved in small quantity by the water; even if a very minute proportion of oxide of lead be dissolved in the water, repeated doses of it will give rise, in the course of time, to the most painful symptoms. Leaden pipes coated internally with tin, to resist the action of water and air, are made by drawing out two concentric cylinders of lead and tin. The use of lead in connection with cider-vats and presses is highly blameable, for the lead is dissolved in large quantity in contact with the acid liquid, and a moderate draught of such cider may easily contain a poisonous dose of the metal.

The want of tenacity exhibited by lead prevents it from being drawn into thin wire.

Its softness enables lead to mark paper, which rubs off minute particles of the metal; the pencils in use for metallic memorandum books are composed of lead hardened by the addition of tin and bismuth.

The easy fusibility of lead adapts it to the use of the type-founder, but it is far too soft to be employed alone for this purpose, and is therefore hardened by the addition of antimony.

Type-metal is an alloy of lead with one-third or one-fourth of its weight of antimony. An improved description of type-metal, lately introduced, is composed of two parts of lead, one part of tin, and one of antimony. Another alloy employed for the same purpose contains fifteen parts of lead, one part of tin, and four parts of antimony.

The high specific gravity as well as the fusibility of lead recommend it for making bullets and small shot, where great momentum is required in a small compass. Rifle bullets must be made of very pure soft lead in order that they may easily take the grooves of the rifle, and the iron projectiles of rifled ordnance are coated with lead for a similar reason, but a somewhat harder lead is employed here; the surface of the shot or shell to be coated is thoroughly cleansed, then dipped into solution of sal-ammoniac, and afterwards into melted zinc, the coating with this metal being found to cause a firmer adhesion of the lead into which the missile is next plunged

Bullets intended to be discharged from smooth-bore small arms, especially those for breech-loaders, which are of small size, are commonly hardened by the addition of one-fifth of their weight of antimony, in order to give them greater penetration. The bullets employed in shrapnel shells are also composed of four parts of lead and one part of antimony, partly for the sake of penetration, and partly that they may scatter better, bullets of soft lead being liable either to be jammed together by the force of the explosion, or so distorted as to make a very short flight when the shell bursts.

Small shot for fowling-pieces are composed of lead containing from three to six parts of arsenic in a thousand, which has the effect, not only of hardening the lead slightly, but also of enabling it to take a nearly spherical form when the melted metal is dropped through a colander into water. Another condition for securing spherical shot is the proper cooling of the drops before they fall into the water, for if they are suddenly chilled and solidified externally long before the inner portion solidifies, the shrinking of the latter, as it cools, causes the outer layer to collapse, and the shot becomes deformed.

Probably the effect of arsenic in securing the spherical shape is due to its diminishing the contraction of the still liquid lead as it cools after the outer portion has solidified.

In order to cool the drops before they enter the water, they are commonly allowed to fall through the air from a considerable height, either in a shot-tower, or in the disused shaft of a mine; or the same object is sometimes attained by employing a rapid blast of air, when a high fall may be dispensed with. The larger the size of the shot, the more preliminary cooling will they require, so that large shot are allowed to fall through 150 feet, and small shot through 100 feet. In order to prepare the metal, it is usual to alloy a quantity of lead with a large amount of arsenic, and to add this to melted lead in the proper proportion. A ton of soft lead is melted in an iron pot, and 40 lbs. of arsenic added to it; the pot is covered with an iron lid, and the joints cemented with clay to prevent the arsenical vapour from escaping; the metal is kept melted for three or four hours, then carefully skimmed, and cast into pigs. The arsenic is added sometimes in the metallic state, sometimes as *white arsenic* (arsenious acid, composed of arsenic and oxygen) or as *orpiment* (sulphuret of arsenic), the two last being decomposed by the lead, and converting a portion of that metal into oxide or sulphuret. Two or three tons of inferior lead having been melted, five or six pigs of the arsenical lead are added, and well stirred up with it; a small sample is allowed to fall from a height, through a perforated ladle, into water; if the drops become flattened, too much arsenic has been added, if they are pear-shaped, there is too little. The colanders are wrought-iron bowls about 10 inches wide, perforated with smooth holes varying in size with the description of shot required. In order somewhat to delay the passage of the melted lead through the holes, the colanders are lined with the *cream* or scum of oxide which forms upon the surface of the metal. The temperature of the lead when poured into the colanders is scarcely adequate to scorch straw. The drops then fall from the top of the shot-tower into a vessel of water at the bottom. They are afterwards dried on a hot plate, sifted into different sizes, the deformed shot rejected by gently shaking on a slightly inclined table, when only the spherical shot roll down, and these are polished in a revolving cask containing a little plumbago.

Lead is extensively employed in the construction of vessels for various chemical manufactures, since it resists the action of sulphuric, muriatic, and fluoric acids in a far higher degree than iron, copper, zinc, or tin. Even nitric acid, if strong, scarcely attacks lead, though the diluted acid readily dissolves it. The large chambers in which sulphuric acid is manufactured are built of leaden plates weighing 5 or 6 lbs. per square foot. Here the fusibility of the metal becomes an advantage, for they have to be united by being *burned together*, that is, by directing a hydrogen flame along the edges of the plates so as to unite them without the intervention of solder, which would soon be corroded under the action of the acid. This is sometimes called *autogenous* soldering.

Notwithstanding that lead is unacted upon in the cold by strong acids, it is very soon extensively corroded when exposed to the action of air in the presence of carbonic acid, and becomes eventually converted into a mass of *white lead* or (basic) carbonate of lead. Since carbonic acid is produced abundantly by the decay and putrefaction of animal and vegetable matters, metallic lead is much affected when kept in contact with such substances in the presence of air, the oxygen of which unites with the lead to produce an oxide of lead which then combines with the carbonic acid and forms a carbonate. The lead of old coffins is sometimes found to have become almost entirely converted into an earthy-looking mass of white lead in this way, a very thin plate of lead remaining in the centre. The oldest process for the manufacture of white lead depends upon the corrosion of the lead in this manner.

In breech-loading cartridges, where grease is employed as a lubricator, the bullets have sometimes become partly converted into white lead, and have thus increased so much in bulk as to burst open the copper case of the cartridge and render it useless.

Alloys of Lead and Tin.—*Pewter** is an alloy of four parts of tin with one part of lead; it is harder, possesses more tenacity, and melts more easily than either of the metals separately, and provided that the lead does not exceed this proportion, the alloy may be used for drinking-vessels without any danger of lead-poisoning. Since, however, lead is far cheaper than tin, a larger proportion than one-fifth of lead is often employed, when the lead is apt to be dissolved if left in contact with the acetic acid always present in beer.

* Probably corrupted from the French, *potée*, pot. *Potée d'étain* is the French for pewter.

Pewter made with the above proportions has the specific gravity 7·8, so that specimens having a higher specific gravity than this will be known to contain more lead.

The solder employed by the pewterer is a very fusible alloy of tin, lead, and bismuth. The solder used for tin-plate is an alloy of lead and tin. *Common solder* contains equal weights of the two metals; *fine solder* contains two parts of tin and one of lead; *coarse solder*, two parts of lead and one of tin. In making solder, the proportions of the metals can be judged of from the appearance of the alloy. When it contains a little more than one-third of its weight of tin, its surface, on cooling, exhibits circular spots due to a partial separation of the metals; but these disappear when the alloy contains two-thirds of its weight of tin. These alloys melt at a much lower temperature than either of their constituent metals. Common solder melts at 385° F.; fine solder at 372° F., whilst the melting-point of tin is 442° F., and that of lead is 620° F.

Soldering is scarcely to be regarded as a merely mechanical adhesion, but depends probably, in part, upon the formation of an alloy between the solder and the surface of the metal to be soldered. Hence it is absolutely necessary that the surfaces to be united by the intervention of solder should be perfectly bright and free from oxide. Several substances are employed to ensure this at the moment of applying the solder; one of the commonest is muriatic (hydrochloric) acid *killed* with zinc, that is, in which a lump of zinc has been dissolved, partly with the object of saturating a portion of the acid, partly to form a chloride of zinc which melts over the surface of the work, dissolving any oxide, and protecting the metal from the oxidising action of the air. Sal-ammoniac (muriate of ammonia), which contains hydrochloric acid combined with ammonia, is also employed, the hydrochloric acid removing any oxide from the metallic surface; sometimes a combination of sal-ammoniac and chloride of zinc is used. Rosin in powder is often sprinkled over the metal to be soldered, when the heat melts it and forms a varnish to protect its surface from the oxygen of the air.

Hard soldering or *brazing*, for uniting the edges of iron, copper, or brass, is effected with an alloy of brass and zinc made by adding zinc to brass melted in a covered crucible; the alloy is granulated by pouring it through a bundle of twigs held over a tub of water, and before being used for brazing it is mixed with a little moistened borax, which melts when the heat is applied, dissolving off any oxide from the metals, and protecting them from the action of the air. For fine work, a little silver is added to the alloy, which is thus rendered much more liquid when fused.

Terne-plate resembles ordinary tin-plate, but is coated with an alloy of tin and lead; it is largely exported to Canada, where it is employed for roofing.

SILVER.

This metal being, in general, a far less chemically active metal than the preceding, that is, being less likely to enter into and remain in a state of chemical combination with other substances, is much more frequently met with in the metallic or native state.

Native silver has generally the appearance of metallic twigs and branches, which are sometimes composed of crystals of silver strung together. The silver-mines of Potosi exhibit such specimens. Native silver is also found at Kongsberg in Norway, at Andreasberg in the Hartz, Freiberg in Saxony, and Schemnitz in Hungary.

The native metal generally contains small quantities of gold and copper. At Kongsberg a yellow alloy is found which contains silver with more than one-fifth of its weight of gold. An amalgam of silver with mercury is found in large quantity in the silver-mines of Coquimbo, Chili.

Sulphuret of silver, or *silver-glance*, containing, in its pure state, 87 parts of silver combined with 13 of sulphur, is one of the commonest forms of combination in which silver occurs in nature. It has been found, in a pure state, in Cornwall, Norway, Hungary, Saxony, Bohemia, Mexico, Peru and the United States. It has a slight lustre, and a dark grey colour, possessing also, which is remarkable in an ore of this description, a good deal of malleability and flexibility, and it is so soft that it may be cut with a knife. It is also distinguished by its fusibility, being easily melted even in an ordinary flame.

The sulphuret of silver is more abundant in association with the sulphurets of other metals; thus, in argentiferous galena, with sulphuret of lead; in grey copper ore, with the sulphurets of copper, antimony, arsenic, iron, and zinc; in *brittle silver ore*, with the sulphurets of antimony, iron, and copper; in *red silver ore*, with sulphuret of antimony or sulphuret of arsenic. Blende, iron pyrites, mispickel, and some other minerals sometimes contain a minute proportion of silver, which may be extracted with profit, incidentally to other processes.

Horn-silver or *chloride of silver* contains, when pure, 75 parts of silver united with 25 parts of chlorine. Good specimens of this ore exhibit, as its name implies, some resemblance to horn in appearance and softness. It is found abundantly in Chili and Peru, sometimes in large fragments, but more commonly in very small cubical crystals disseminated in a ferruginous rock.

The *butter-milk ore* of the German miners contains chloride of silver mixed with a large proportion of clay.

Chloride of silver has been found, in small quantity, in Cornwall. *Bromide* and *iodide of silver* are also found in Mexico and Chili.

In consequence of the high price of silver, it admits of being extracted with profit even from ores which contain a very small proportion of the metal, especially if some other useful metal can be extracted at the same time. Thus, silver may be profitably obtained from galena containing only two parts of silver in a thousand, and even a smaller quantity than this is extracted from some copper ores. In these cases, the lead and copper respectively are extracted by the ordinary smelting processes, and are then subjected to special treatment for the extraction of the silver. It may be well to describe the principal methods in use for this purpose before considering the metallurgic treatment of the ores of silver with the sole object of extracting that metal.

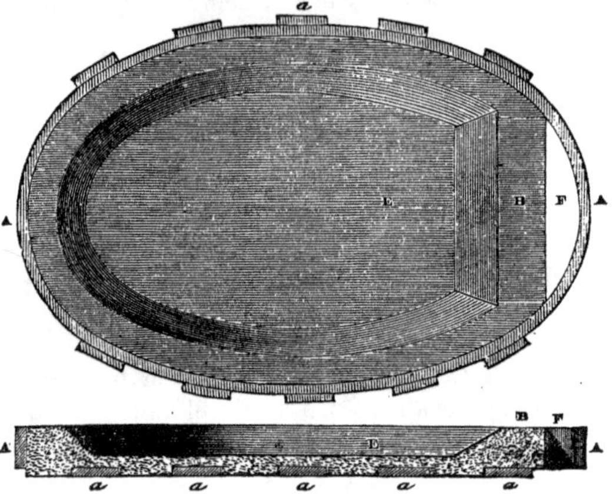

FIG. 74.—English Cupel or Test.

Extraction of Silver from Lead by Cupellation.—Most of the silver produced in this country is extracted by this process from the rich lead obtained in Pattinson's desilverising process (p. 194).

The process of cupellation, which is one of the most attractive metallurgic operations, derives its name either from the German *kuppel*, a *cupola* or *dome*, in allusion to the shape of the German cupellation furnace, or from a diminutive derived from the Latin *cupa*, a cup, referring to the concave hearth upon which the process is carried out.

The extraction of silver from lead by cupellation depends upon the facility with which the latter metal is converted into an oxide by the action of air at a high temperature, whilst silver is almost entirely unaffected; the oxide of lead being easily melted, is partly removed from the surface in a liquid state, and partly absorbed by the porous hearth upon which the silver remains. In England, this hearth, which is called the *cupel* or *test*, is an oval frame of wrought iron (A, Fig. 74), 5 feet long and $2\frac{1}{2}$ feet wide, which is crossed by five iron bars (*a*) $3\frac{3}{4}$ inches in breadth. This frame is filled with finely-powdered bone ashes moistened with water, in which a little pearl-ash (carbonate of potash) has been dissolved; this is well consolidated by beating, and scooped out until it is about $\frac{3}{4}$ of an inch thick over the cross bars, leaving a flat rim of bone-ash all round, about 2 inches wide, except at one end (B), the front or *breast* of the cupel, where it is 5 inches wide; through this a channel (F) is cut, to allow the melted oxide of lead to flow off without coming in contact with the iron, which it would corrode very seriously. This cupel rests upon a car, in order that it may be wheeled into its place under the reverberatory furnace (Fig. 75), of which

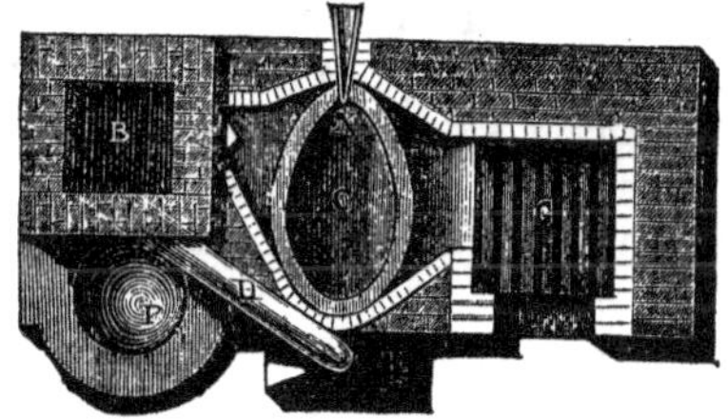

FIG. 75.—English Furnace for the cupellation of Lead.

it forms the hearth (C), where it is arranged so that the flame of a coal fire (G) passes directly across it to the two flues, which run into a chimney (B), about 40 feet high. A blast pipe or tuyère (N) enters the cupel at the end opposite to that through which the oxide of lead escapes, and throws a blast of air over the surface of the metal at the rate of about 200 cubic feet per minute. Opposite to the tuyère is a hood with a pipe (H) for carrying the fumes into the chimney. The cupel is very gradually heated nearly to redness, and almost filled with the lead to be treated, which is ladled in from an iron pot (P), in which it has been previously melted. From 500 to 600 lbs. of lead are introduced at once. In a short time, the surface of the metal becomes covered with oxide of lead, or *litharge*,* in a melted state; the blast is then turned on, and drives the litharge in wavelets off the surface, through the channel made for its escape, into a cast-iron pot outside the furnace. When this channel is very much corroded, it is closed, and another is cut. In proportion as the lead upon the hearth diminishes, fresh portions are ladled in from the melting-pot, so as to keep the lead at about the same level. The process is continued for sixteen or eighteen hours, in which time four or five tons of lead will have been added, and an alloy containing about eight parts of silver in a hundred will be left in the cupel. A hole is then made in the bottom, through which the metal is run out and cast into pigs. A fresh charge is introduced into the cupel, and the operation continued; one cupel will often last for forty-eight hours, and is capable of treating 5 cwts. of lead per hour, with a consumption of 1 cwt. of coal. When about 3 tons of the rich alloy (containing 8 per cent. of silver) have been obtained, it is again subjected to cupellation in the same furnace, but in a cupel which has a concavity at the bottom intended for the reception of the cake of silver, which will weigh about 500 lbs. This second operation is

* From two Greek words, signifying *stone* and *silver*.

necessary, because the litharge which is formed from this very rich alloy contains a considerable proportion of silver, so that the lead obtained by smelting it (p. 212) and by smelting the cupel, which absorbs a large proportion of litharge, contains 30 or 40 ozs. of silver in the ton, and must be treated for that metal.

The appearance of the metal in the cupel as the last portions of lead are removed in the form of litharge and absorbed into the bone-ash, is very beautiful. When the film of oxide becomes so thin that the bright silver beneath can reflect the light through it, a decomposition of the light into its constituent colours takes place, and the most brilliant rainbow tints are seen upon the metal, their beauty being enhanced by the rapid rotation of the film. As soon as this film of oxide has been absorbed by the cupel, the splendid surface of the melted silver shines out (*fulguration*, *coruscation*, or *brightening*).

During the cooling of the cake of silver, some very remarkable phenomena are observed. When a thin crust of metal has formed upon the surface, the silver beneath it assumes the appearance of boiling, and the crust is forced up into hollow cones about an inch high, through which the melted silver is thrown out with explosive violence, some of it being splashed against the arch of the furnace, and some solidifying into most fantastic tree-like forms several inches in height. This behaviour of silver has been shown to be due to its property of absorbing mechanically (*occluding*) oxygen, at a temperature above its melting-point, which it gives off as it approaches the point of solidification, the escaping gas forcing up the crust of solid silver formed upon the surface.

A considerable proportion of lead and silver is carried off by the blast, in the form of vapour, and is partially recovered, as oxide, from the flues of the furnace.

In some cupellation furnaces, instead of employing a blowing machine, a current of air is directed over the surface by means of a jet of steam issuing from a tube surrounded by a wider one through which the air is dragged by the mechanical action of the steam. This is said to hasten the operation, and to produce litharge of better quality.

The litharge produced in the English cupellation furnace is reduced to the metallic state in a reverberatory furnace with a hearth measuring 8 feet by 7, which is lined with bituminous coal; this soon becomes converted into a porous coke, which protects the clay hearth of the furnace from

FIG. 76—German Cupellation-furnace.

being corroded by the melted litharge, and forms a filter through which the lead runs towards the opening from which it is tapped. About 3 tons of litharge mixed with 6 cwts. of small coal are charged at once; the carbon of the coal removes the oxygen from the oxide of lead composing the litharge, and reduces the lead to the metallic state. The lead thus obtained usually contains 30 or 40 ozs. of silver in a ton, and is introduced into its appropriate place in a series of Pattinson's pans (p. 194).

The *German cupellation-furnace* (Figs. 76, 77) differs from the English in having a fixed instead of a moveable hearth (A), covered with an iron dome (C) lined with clay, which is capable of being lifted off by a crane (G). The hearth is circular, about 10 feet wide, and is lined with marl, or with an intimate mixture of clay and lime well beaten and hollowed out like a saucer, with a circular cavity about 20 inches wide and ½ inch deep, in the middle, for collecting the cake of silver. Wood ashes, previously washed and well beaten down, are sometimes employed instead of marl. Two tuyères (*a*) direct the blast across the hearth, and are pro-

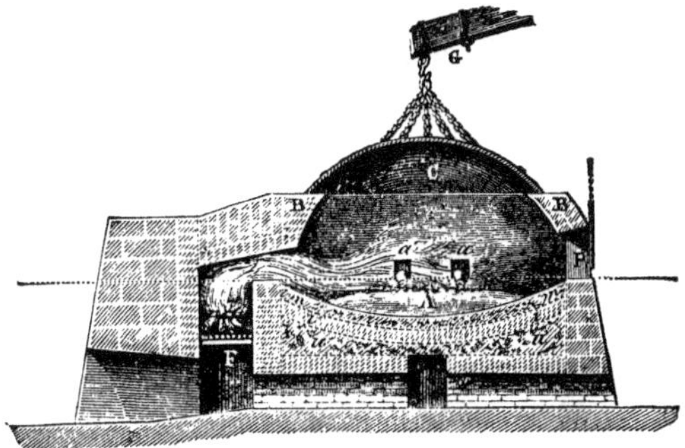

FIG. 77.—Section of German Cupellation-furnace.

vided with *butterfly-valves* for guiding the blast over the surface of the metal. The fire-place (F) is situated in a square furnace adjoining the hearth, and is supplied, when practicable, with wood, which gives a longer and clearer flame than coal. Impure leads can be treated by the German process, whilst the English method of cupellation is adapted for those which are comparatively free from antimony and copper.

From 4 to 17 tons of lead can be cupelled at once, according to the size of the hearth, the pigs being placed upon a thin layer of straw. The heat is gradually raised so as to melt the lead, no blast being employed for the first three hours. When the metal is in a state of tranquil fusion, the surface is skimmed to remove the dross, and the bellows are worked at the rate of about four or five strokes a minute, in order to renew the air over the surface, so as to promote oxidation. In about two hours, a stronger fire is applied, and the crust of oxide and of various mechanical impurities is skimmed off the surface through the opening (*o*) provided for the escape of the litharge. About an hour and a half is occupied in thoroughly cleansing the surface of the lead.

The blast is now freely directed upon the melted metal, so as to produce litharge abundantly, and to drive it in waves through the outlet (*o*), which is deepened by the workman in proportion as the level of the lead falls. The litharge flows out on to the floor of the shop as at L. After continuing the process for a period varying between seven hours and sixty hours according to the amount of lead, the removal of the lead is complete so far as it can be effected in this furnace, and the phenomena described at p. 211 as indicating the termination, are witnessed. A wooden spout is then introduced into the charging-door, through which water is carefully poured upon the surface of the silver to solidify it into a cake, which undergoes a subsequent refining to complete its purification.

The first portions of oxide which form upon the surface in this process contain, beside the oxide of lead, oxides of antimony, iron and other impurities, and yield, when reduced, a very impure lead, fit only for making shot or type-metal. The last portions of litharge, amounting to about $\frac{1}{20}$th of the whole, are set aside to be reduced separately, when they furnish a lead rich in silver. When bismuth is present in the lead, the last portions of litharge have a green colour. A certain quantity of the intermediate portion of litharge (containing less than half an ounce of silver in the ton) is sent into the market as such, being useful to the manufacturers of glass and earthenware, of sugar of lead, &c. In the market, it finds a readier sale when in reddish brown scales or flakes, which are produced by running it out, when melted, into large iron vessels, and allowing it to cool in a draught of air. When these vessels are inverted, the mass of litharge is easily turned out and broken into a flaky powder. Impure litharge cannot be converted into red litharge, so that the colour is evidence of its purity.

The cupel, which is largely impregnated with litharge and contains some silver, is broken up and smelted in order to extract those metals.

At Andreasberg and Freiberg, the silver ores are melted with the lead (already containing some silver) upon the cupel itself, when any sulphuret of silver which they contain is decomposed, its sulphur combining with oxygen to form sulphurous acid, and the silver being dissolved by the lead.

At Kongsberg, the use of a blast heated to about 400° F. has been attended with a great saving of time and fuel, with the additional advantage of cupelling leads containing ten or twelve parts of copper in the hundred, which are too difficult to fuse on the cold blast cupel. In a furnace six feet in diameter, a charge of 3½ tons of lead is cupelled in seven hours.

The silver obtained by cupellation is liable to contain small quantities of lead, bismuth, antimony, copper and gold, the three first of which render it brittle. It is therefore generally subjected to a refining process, which consists in exposing it in a melted state to the action of the air, when the foreign metals, with the exception of the gold, are oxidised and converted into a dross. The operation is performed either in an ordinary cupellation furnace, or in another constructed upon the same principle.

Extraction of Silver from the Ores of Copper.—The ores of copper containing silver are smelted chiefly at Mansfeld, the copper being extracted in the form of *black copper* by the process described at p. 125. From the black copper the silver is extracted by the process of *eliquation*, which consists in alloying it with a large proportion of lead, and afterwards melting out the latter metal by a moderate heat, when it carries all the silver with it, to be afterwards extracted by cupellation. The black copper is broken into small fragments, or granulated by melting it and running it into water, and is then fused in a small blast furnace with from two to four times its weight of lead, that which already contains silver being chosen if possible, and the two metals thrown in alternately. The alloy of lead and copper is cooled quickly in thick cast-iron moulds, and chilled with water in order to avoid the separation of the two metals, being thus cast into round cakes 18 inches in diameter and 3 inches thick. These cakes are placed on a *liquation-hearth* (Fig. 78), which is a gutter made by two sloping cast-iron plates with a space between them through which the lead trickles down. The cakes of copper-lead (D) are set on

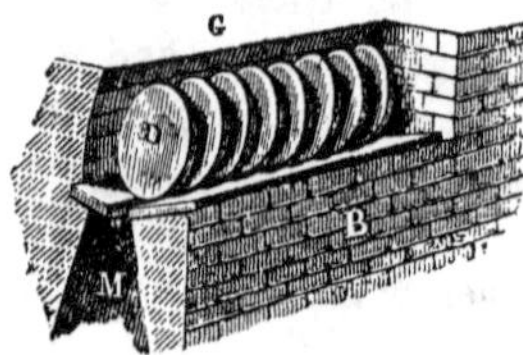

FIG. 78.—Liquation-hearth.

FIG. 79.—Liquation-hearth.

edge across this gutter, with pieces of wood to keep them apart; the gutter is then shut in with iron plates (F, Fig. 79), and filled with charcoal. A wood fire being made in the space (M) beneath the gutter, the charcoal takes fire and melts the lead, which runs into a receptacle (*o*) outside the furnace, carrying the silver with it. This lead is cast into ingots from which the silver is extracted by cupellation. The liquation occupies three or four hours. The copper cakes still retain about one-fourth of lead and some silver, which are extracted by exposing them to a higher temperature in a *sweating-furnace* (Fig. 80) where they are placed over a number of fire-brick channels (F) in which a wood fire is made. The bulk of the lead is converted into oxide by the air in the furnace, so that a quantity of litharge containing oxide of copper and silver collects at the bottom of the channels, and is fused with the black copper in the blast furnace employed for preparing the cakes to be submitted to eliquation.

FIG. 80.—Sweating-furnace. M, Back-wall with flues communicating with the chimney H.

Extraction of Silver from the Ore by Melting with Lead.—At Kongsberg in Norway, where the ore contains its silver in the metallic state, it is extracted by simply melting the dressed ore with its own weight of lead, when an alloy containing about one-third of its weight of silver is obtained, which is submitted to the process of cupellation. When ores containing sulphuret of silver are melted with lead, this metal removes the sulphur in the form of sulphuret of lead, and the liberated silver is dissolved by another portion of lead.

AMALGAMATION PROCESS FOR THE EXTRACTION OF SILVER FROM ITS ORES.

The process of amalgamation is so called because the silver is extracted in the form of an *amalgam* with mercury, and there are two modes of carrying it out, the choice being regulated by the scarcity or abundance of fuel in the locality. Thus, the amalgamation process employed in Mexico and Chili, where fuel is dear, is very different from that in use in Freiberg where it may be had in abundance.

Mexican Process of Amalgamation.—In Mexico and Chili, where the silver ores obtained from the western slopes of the Cordilleras are treated, the thorough pounding of the ores and of the materials added to them, by the comparatively inexpensive agency of water-power, horses and mules, is employed to facilitate the chemical changes which would be promoted by the action of fire, if fuel were more abundant. Time is also an important element in this process, which consists essentially in converting the whole of the silver into chloride of silver, to be afterwards brought to the metallic state by the action of mercury, and dissolved by the latter, in the form of an amalgam.

The ore sometimes contains only 35 ozs. of silver in the ton, partly in the metallic state, and partly as chloride and sulphuret of silver, and is carefully picked over in order that the worthless portions may be rejected. It is then crushed in oblong mortars under wooden pestles, shod with iron, weighing about 200 lbs. each, and raised by cams projecting from an axle moved by a water-wheel. The crushed ore is ground with water into a mud, under granite stones, which are made to revolve in a granite bed by the labour of mules. This mud is transferred to the *amalgamation-floor*, an enclosure, about 300 feet by 240 feet, paved with stone. It is there mixed, with wooden shovels, with a proportion of common salt, varying from one to five parts for every hundred parts of ore, and, in order to effect a thorough intermixture, a number of horses are made to trample the mud in the enclosure, after which it is left at rest for some hours.

On the following morning, the horses are again turned into the amalgamation-floor for an hour, after which, copper pyrites which has been roasted and ground to powder (when it is called *magistral*) is added in the proportion of $\frac{1}{50}$th or $\frac{1}{100}$th of the weight of the ore. Six horses attached to a halter held by a man in the centre, are then made to trample the mud for five or six hours.

The magistral contains about $\frac{1}{10}$th of its weight of sulphate of copper, which decomposes the common salt (chloride of sodium), yielding sulphate of sodium and chloride of copper.

The mixture is now ready for the amalgamation with mercury, which is sprinkled upon it from a bag of coarse canvas, in quantity about twice that of the silver present in the ore. It is then trodden again by horses, and well turned with wooden shovels, these operations being repeated every other day until, on washing a small sample in a bowl, it is found that all the mercury which has been added is in combination with the silver, as an amalgam, no globules of mercury being visible.

The chloride of silver has been decomposed by the mercury, forming chloride of mercury (*calomel*) and metallic silver, the latter having combined with the mercury to form an *amalgam of silver*.

The chloride of copper, exposed to the action of air and moisture, acts upon the sulphuret of silver, the sulphur of which is set free, the silver being converted into chloride by removing half the chlorine from the chloride of copper, which is then converted into *oxychloride of copper* by absorbing oxygen from the air. The chloride of silver is decomposed, as above stated, by the mercury, with formation of calomel and amalgam of silver.*

The examination of the amalgam shows whether too much sulphate of copper (in the form of *magistral*) has been added. If this be the case, a quantity of the mercury will be found to be so finely divided as to form a dark-coloured

* For the explanation of this part of the amalgamation process, the author is indebted to the kindness of Dr. Percy.

mud, and occasional brown spots of metallic copper will be visible. A great loss of mercury would ensue if the magistral had been added in excess, because the chloride of copper formed from it would be decomposed by the mercury, yielding a chloride of mercury (calomel) and subchloride of copper.

If it is found that this error has been committed, a little lime is added to the contents of the amalgamation-floor to decompose the chloride of copper (forming chloride of calcium and oxide of copper) and prevent any further waste of mercury.

It sometimes occurs, on washing the sample, that globules of mercury are perceived, showing that it has not entirely united with the silver. This is due to an insufficient supply of magistral, in consequence of which the amount of chloride of copper formed has not been sufficient to convert the whole of the silver into chloride of silver, a form in which it is much more readily acted upon by the mercury than when it is in the state of sulphuret of silver, because the chloride of silver is capable of being dissolved by the strong solution of common salt existing in the mud, and is thus presented to the mercury in the condition most favourable to chemical action. If necessary, a further addition of the roasted copper pyrites is made before proceeding with the amalgamation.

After about a fortnight, a fresh quantity of mercury is added, rather less than one-third of the first addition, and the operation of trampling is repeated. When this mercury has also been taken up, a third portion, about half as large again as the second, is added, in order to dissolve the amalgam of silver in an excess of mercury. After a thorough incorporation by trampling, the mixture is at once shovelled into barrows and taken to the washing-vat, which is a circular cistern 8 feet wide and 9 feet deep, in which the mud from the amalgamating-floor is well stirred up with water, by an agitator worked by four mules, fresh water constantly running into the cistern, and carrying off the earthy matter over the side.

The liquid amalgam left behind is thrown into a leather bag with a canvas bottom, through which the excess of mercury is strained off, leaving a pasty amalgam containing about $\frac{1}{5}$th or $\frac{1}{6}$th of its weight of silver, which is subjected to pressure in order to squeeze out more mercury and convert it into a hard solid mass. This is moulded into wedge-shaped masses of about 30 lbs. each, which are built up into a circular tower on a copper stand having a hole in its centre through which a pipe passes in order to conduct the mercurial vapours into a tank of water placed beneath. The pile of amalgam is covered with an iron bell, the opening of which fits tightly upon the copper stand, any crevices being carefully filled up with a cement. A temporary furnace is built with bricks around the bell, and kept full of burning charcoal for about twenty hours, when all the mercury is converted into vapour, which condenses in the water beneath, and the silver is left as a hard mass; this is broken up, melted, and cast into bars weighing about 1000 ozs. each.

The period occupied by the amalgamation process varies greatly according to the nature of the ores, extending sometimes over little more than a fortnight, and sometimes requiring six or eight weeks.

The loss of mercury in the process is very considerable, amounting to about 24 ozs. for every pound of silver obtained, being due, in great part, to the *flouring* or fine division of the metal, which is then carried away in the process of washing, together with all the silver which has been dissolved in it.

The *hot amalgamation*, applied for extracting the silver from rich ores which contain it either in the metallic state or as chloride, bromide or iodide, only occupies five or six hours, and entails less loss of mercury. It consists in boiling the finely-powdered ore with water, common salt (10 or 15 parts for a hundred of ore) and mercury, in pans with copper bottoms, when the chloride, bromide and iodide of silver are decomposed by the copper, forming a chloride, bromide or iodide of copper, and liberating the silver, which is dissolved by the mercury.

The extraction of silver (and of gold) by amalgamation is very much facilitated by adding a minute proportion of *sodium* to the mercury employed, for it is found that mercury so treated attacks and dissolves silver and gold much more readily than pure mercury, and that it is very much less liable to assume that finely divided state in which it is so readily washed away and lost.

Since sodium is rapidly oxidised by exposure to air, and decomposes water with explosive violence, it is not a very portable substance; hence it is better that it should be carried in the form of a strong *amalgam of sodium* which may be diluted with the proper quantity of mercury on the spot where it is to be used.

The strong amalgam is prepared by heating 100 parts of mercury to about 300° F., in an iron vessel with a narrow neck, and adding, in small pieces (since the combination is very violent), 15 parts of metallic sodium; the amalgam is then poured out into ingot moulds, where it solidifies into very brittle crystalline bars. One part of this amalgam is melted by a gentle heat and mixed with about five thousand parts of mercury before employing it in the amalgamating process. The action of the sodium appears to depend upon its highly *electropositive* character, but its mode of operation does not appear to be as yet well understood.

Amalgamation Process for Extraction of Silver at Freiberg.—The process adopted for extracting the silver from the ores mined in Saxony is far less wasteful of mercury than the Mexican process just described, the loss varying from 4 to 12 ozs. per lb. of silver, since this valuable metal is employed only for the purpose of dissolving the silver after it has been brought into the metallic state, whereas in Mexico, a part of the mercury is used in removing the chlorine from the chloride of silver in order to liberate the metal. In the Saxon process, as in the Mexican, the whole of the silver is converted into chloride of silver, from which the chlorine is removed by the action of metallic iron, and the silver, which has thus been reduced to the metallic state, is dissolved by mercury.

The ore treated at Freiberg contains the silver chiefly as a sulphuret, but it also contains the sulphurets of several other metals, particularly of antimony, bismuth, arsenic, iron, copper, lead and zinc. The different kinds of ore are sorted and mixed so that the mixture may contain less than one part of copper and five parts of lead in the hundred parts, for both these are easily taken up by the mercury, causing undue consumption of that metal. The silver present in the mixed ores should amount to about 80 ozs. in the ton. The presence of a large proportion of iron pyrites (bisulphuret of iron) is necessary for the subsequent chemical changes, so that the mixture is made to contain about one-third of its weight of that mineral.

The ore thus prepared is ground to a coarse powder, mixed with one-tenth of its weight of common salt, and roasted upon the hearth of a reverberatory furnace at a dull red heat, for about two hours, care being taken to turn it over frequently, and to avoid fusion. During the roasting, the oxygen of the air converts the arsenic and antimony into their respective oxides, which pass off as a thick white smoke, together with some sulphurous acid produced by the combustion of the sulphur. The sulphurets of copper, iron and silver, also combine with oxygen, and become converted into the sulphates of those three metals.

The temperature is then raised to enable the sulphates of copper, iron and silver to decompose the chloride of sodium (common salt) yielding sulphate of sodium, and chlorides of copper, iron and silver. Other chemical changes also occur in this stage, but it will not be necessary to trace them, since they do not affect the extraction of the silver.

FIG. 81.—Amalgamation Cask. *a*, Opening for charging and discharging. *r r''*, Toothed wheel for receiving motion from the axis (Fig. 82).

The first roasting is effected with a coal fire, and the second heating with fir-wood.

The deep brown roasted ore, containing the sulphate of silver, is now sifted, and the lumps mixed with more salt, and roasted again in order to complete the change.

The sifted powder is ground to a very fine meal, and introduced in charges of half a ton into strong oaken casks (Fig. 81), about three feet each way, containing three hundred weights of water, and mounted, to the number of twenty, on cast-iron axles, so that they can be made to revolve at pleasure by being connected with the axle (A, Fig. 82) of a water-wheel. About one hundred weight of wrought iron, in fragments, is put into each of the casks, which are then made to revolve at the rate of ten or twelve turns in a minute. The iron removes the chlorine from the chloride of silver, producing chloride of iron and metallic silver, which remains dispersed through the mixture in a finely-divided condition.

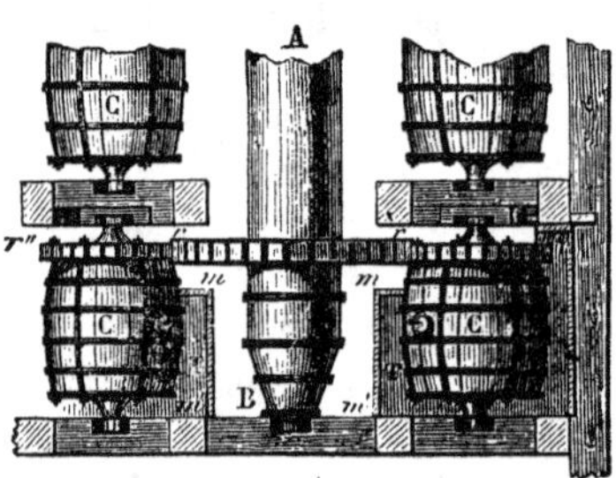

FIG. 82.—Amalgamation of Silver Ores at Freiberg. *r r''*, Toothed wheels for transmitting motion from the axis A B. *m m'*, Troughs for receiving the contents of the casks C.

The iron also decomposes the chloride of copper in a similar manner, yielding finely divided metallic copper.

After having revolved for about two hours, the casks are opened for the purpose of introducing the mercury to dissolve the silver; but before adding this, the contents must be brought to a proper consistence; if they are too thick, they could not be well mixed with the mercury, and some more water must be added; on the other hand, if they are very liquid, the mercury will at once go to the bottom, and will never be thoroughly distributed through the mixture; some more of the prepared ore must then be introduced. 5 cwts. of mercury are then poured into each of the casks, and they are made to revolve, at the rate of about twenty-five turns in a minute, for sixteen or eighteen hours, when the mercury dissolves the metallic silver and copper. During this process, the contents of the casks are twice examined in order to see if they have a proper consistence. The chemical action which takes place in the casks raises their temperature, even in winter, to about 100° F.

In a more modern arrangement, the casks are fixed in a vertical position, the materials being mixed by a revolving agitator with iron arms. Steam is admitted into the casks through holes at the bottom, in order to facilitate the amalgamation by its heat.

When this rotation is finished, the amalgam of silver and copper is found interspersed in minute globules throughout the mass; in order to collect it, the casks are filled with water, and revolved eight times in a minute, during about two hours. They are then turned with their wooden bungs (*a*, Fig. 83) downwards, and the pegs stopping the small openings in the bungs being withdrawn, a tube with a stopcock is inserted; so that the amalgam may run out into a trough beneath. The mixture in the casks is emptied out through a grating which retains the pieces of iron, whilst the mud is collected in tubs, where it is stirred with water and allowed to deposit the small quantity of amalgam which has been carried away.

The mud is afterwards allowed to settle down so that it may be again treated by the amalgamation process, when it yields about 4½ ounces of silver for a ton.

The liquid amalgam of silver and copper is strained through canvas bags, as in the Mexican process, when a quantity of liquid mercury runs off (containing about 20 ozs. of silver to the ton), leaving in the bags a pasty amalgam containing about 30 parts of mercury, 4 parts of silver, and

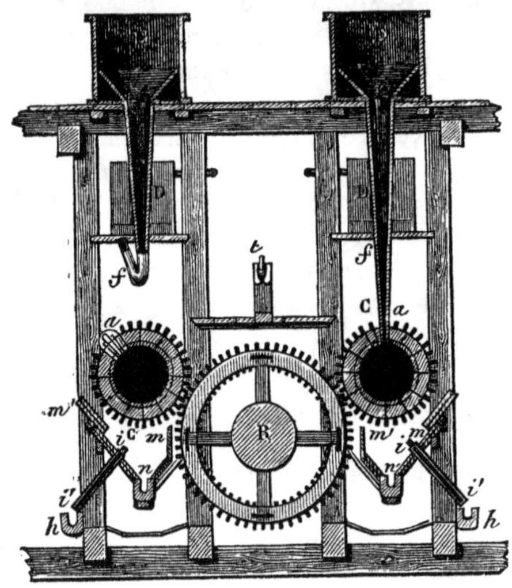

FIG. 83.—Amalgamation of Silver Ores at Freiberg. E, Reservoirs for the ground mineral. *f*, Leathern pipe for conveying it into the opening *a* of the cask C. D, Cisterns containing water. *m n m'*, Receptacles for the amalgam and spent charge. *i i'*, Pipe for conveying the amalgam into the gutter *h*.

1 part of copper, with small quantities of other metals which existed in the ore, particularly lead, bismuth, zinc, antimony, and gold. The amalgam is sometimes allowed to settle in a narrow wooden cylinder, 8 feet high, before being strained in the bags, when it separates into two layers, the lower consisting chiefly of mercury, which need not be strained.

The pressure is sometimes applied to the amalgam by a

screw, and sometimes by the hydraulic press; in the latter case the amalgam is placed in an iron cylinder with a wooden bottom through which the liquid mercury is pressed out.

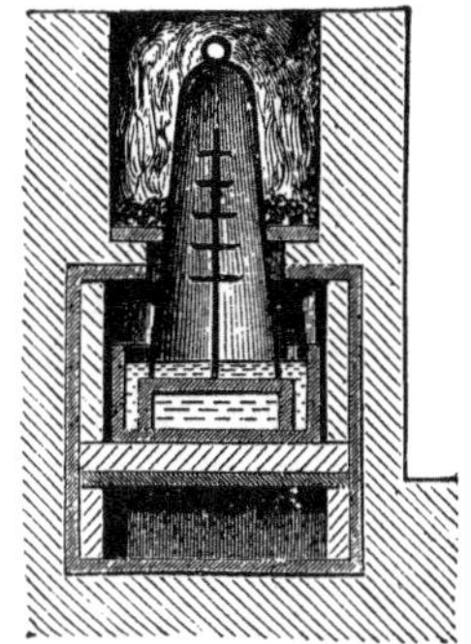

FIG. 84.—Distillation of the Amalgam of Silver.

This pasty amalgam is made up into balls which are placed in iron dishes (Fig. 84) supported at about five inches apart by an iron rod which runs through their centres, and stands, upon four feet, in an iron basin of water. When about 3 cwts. of amalgam have been placed in the dishes, an iron bell is let down over them, by a crane, its lower opening resting in the water. A fire is then made around the upper part of the bell, first with wood, then with turf, and finally with charcoal. In this way the bell is made red hot, and the mercury is converted into vapour, condensing in the water beneath, which is kept constantly cool by the passage of cold water through a wooden trough in which the iron

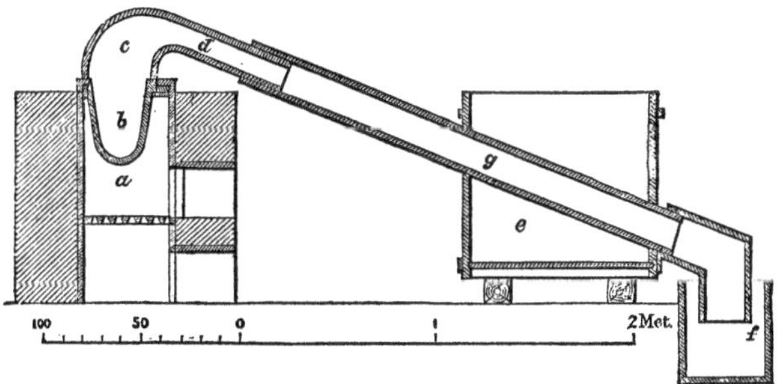

FIG. 85.—Apparatus for distillation of the Amalgam of Silver.

basin stands. After about eight hours, no more globules of mercury are heard to drop into the water, and the distillation is finished. On removing the bell, spongy masses of the alloy of silver, copper, &c., are found in the iron dishes.

A more modern and economical apparatus for the distillation of the pasty amalgam consists of an iron crucible (*b*, Fig. 85) 22 inches wide and 11 inches deep, which is heated by a charcoal fire (*a*), the vapour of mercury being conducted by an iron hood (*c*) into an iron condensing tube (*g*) which traverses a cistern of water (*e*). The inside of the crucible is coated with lime, and an iron plate coated with lime is fitted into the crucible, which is withdrawn with the silver adhering to it at the close of the operation. 4 cwts. of amalgam are distilled in five hours.

The alloy left after distilling the amalgam is melted in crucibles made of a mixture of fire-clay and plumbago, and briskly stirred with an iron rod. Fumes of the oxides of bismuth, zinc and antimony are given off, and a scum containing oxides of lead and copper forms upon the surface, and is skimmed off. When no more dross appears upon the surface, the metal is cast into ingots. It still contains about one-seventh of its weight of copper, from which it is purified by melting it with lead, and subjecting it to cupellation (p. 208), when the litharge dissolves the copper in the form of oxide, and the silver is left pure.

Treatment of Copper-matts for Silver.—When the last matts (corresponding to the Welsh fine metal) obtained in the copper-smelting at Mansfeld (p. 125) contain any considerable proportion of silver (not less than 2½ parts in a thousand), it is extracted by a process closely resembling that employed at Freiberg. Since, however, these matts consist almost entirely of sulphuret of copper and sulphuret of iron, so large a quantity of the sulphates of these metals is formed during the roasting, that it is found necessary to mix the roasted ore with a quantity of chalk equal to that of the common salt; when the lime in the chalk decomposes the sulphates of iron and copper, forming sulphate of lime, and leaving the iron and copper in the form of oxides.

The mud emptied from the tubs at the close of the amalgamation process, at Mansfeld, is mixed with clay, and made up into cakes, which are dried and smelted in a small blast-furnace to obtain the copper.

Processes employed to supersede the Amalgamation of Silver Ores. (*Augustin's Process.*)—Several methods have been proposed from time to time to avoid the use of mercury in extracting silver from its ores. For example, after roasting the ores, first by themselves, and afterwards in admixture with common salt, as in the Freiberg process, to convert the whole of the silver into chloride, the latter is dissolved out with a saturated solution of common salt, and the chloride of silver in the solution is decomposed by leaving it in contact with scraps of copper, when the latter combines with the chlorine, forming a chloride of copper, and the silver is separated in the finely-divided metallic state, to be afterwards melted and cast into ingots. It has been recommended to mix chlorine-water with the solution of common salt employed in this process, in order to extract the gold at the same time.

This process has been employed with satisfactory results, at Freiberg, for the extraction of silver from the matt containing two-thirds of its weight of copper obtained in the copper-smelting process. The matt, having been roasted by itself, and afterwards with common salt, is stamped to powder, sifted, and placed, in quantities of about 6 cwts., in tubs with perforated bottoms, over which a linen strainer is stretched. A hot strong solution of salt being let in from a reservoir, dissolves the chloride of silver and carries it through the strainer into tubs also provided with a strainer, and containing some spongy copper (cement copper, see p. 128), which is gradually dissolved, as chloride of copper, whilst metallic silver is deposited in its stead; this is removed from the tubs, washed with muriatic acid to remove particles of copper, then with water, and moulded into small balls, which are dried and fused.

The liquid containing the chloride of copper and common salt is conveyed into another series of tubs containing copper, in order to ensure the complete removal of the silver, and afterwards into vessels containing iron, which causes the separation of the metallic copper, to be used over again for precipitating the silver.

The matt, having been washed with solution of salt, until a copper plate dipped into the liquor is no longer whitened by the deposition of silver, is washed with water, and taken to the copper-smelting furnace (p. 125).

Patera's process, as carried out at Joachimstal, is applicable only to rich ores, and consists in roasting with common salt, as at Freiberg, in order to convert the silver into chloride of silver, which is then dissolved out by a cold dilute solution of *hyposulphite of soda*, which takes up chloride of silver much more readily than common salt does. The solution is then mixed with a solution of *sulphuret of sodium*, which produces a black precipitate of sulphuret of silver, and leaves the solution of hyposulphite ready to be employed for treating a fresh portion of ore. The sulphuret of silver is collected on a canvas strainer, washed, dried, roasted to burn off part of the sulphur, and melted in black lead crucibles with metallic iron, which takes up the rest of the sulphur, in the form of sulphuret of iron, and leaves metallic silver. The sulphuret of iron dissolves a part of the silver, and is worked up with a fresh charge of ore.

Ziervogel's Process.—A still simpler process consists in roasting the ore or matt containing the sulphurets of copper, iron and silver, so that these may combine with oxygen from the air, and become converted into sulphates, which are then heated so strongly as to decompose the sulphates of copper and iron, leaving these metals as insoluble oxides, whilst the sulphate of silver is left undecomposed, and may be dissolved out by water and placed in contact with metallic copper, which separates the silver in the metallic state, and takes its place in the solution, yielding sulphate of copper.

Ziervogel's process has been attended with very satisfactory results in its application to the extraction of silver from the copper-matts at Mansfeld, containing, in 100 parts, 80 parts of subsulphuret of copper, 11 parts of sulphuret of iron, and $\frac{3}{8}$ths part of sulphuret of silver. The loss of silver experienced in treating this matt by the amalgamation process amounted to nearly $\frac{1}{10}$th of the metal present, by Augustin's process to nearly $\frac{1}{12}$th, and by Ziervogel's process to only $\frac{1}{120}$th.

The matt, having been granulated, ground, and sifted, is roasted with great care and judgment, in reverberatory furnaces, seven of which are connected with a single chimney 154 feet high. The roasting occupies about ten hours, and is continued until a small sample taken out and mixed with water gives a liquor which produces a strong precipitate of chloride of silver on the addition of common salt (chloride of sodium), and has a light blue colour, from a little sulphate of copper having been left undecomposed.

The roasted mass is then introduced, in quantities of 5 cwts., into tubs similar to those used in Augustin's process (p. 229), and treated with hot water containing a little sulphuric acid, until the liquor which runs off no longer becomes milky when mixed with common salt, showing it to be free from silver. The solution of sulphate of silver, thus obtained, is run into tubs containing copper, where the silver is precipitated, and is afterwards washed with diluted sulphuric acid to remove adhering copper.

The sulphate of copper in solution is decomposed by metallic iron, and the copper is employed for precipitating the silver.

The copper matt of Freiberg is sometimes roasted, to convert the sulphuret of copper into oxide, and is then boiled in leaden tubs with diluted sulphuric acid, which dissolves the oxide of copper in the form of sulphate of copper; this salt is obtained in marketable crystals from the solution. The residue, which contains the silver and gold, is washed, mixed with half its weight of litharge, and made up into balls, which are dried and smelted together with more litharge and lead-slags, when the lead obtained contains all the silver and gold, which are recovered from it by cupellation.

Very poor ores have sometimes been treated by melting them, either in cupola or reverberatory furnaces, with iron pyrites, which takes up the silver. By smelting this pyrites together with galena, the silver is obtained with the lead, which may be separated from it as usual (p. 194).

A considerable quantity of silver is now extracted from the Wicklow pyrites (sold as *silver smalls*), after it has been calcined for its sulphur, in the manufacture of sulphuric acid.

Applications of Silver.—Pure silver is far too soft to resist the wear to which it is subjected in common use. It is therefore hardened by alloying it with copper, a considerable quantity of which may be added without material alteration of colour. The hardest alloy is that which contains 4 parts of silver and 1 part of copper. The standard silver used for coin and for silver articles in England contains, in 100 parts, 92½ parts of silver and 7½ parts of copper; the French silver coinage contains 90 parts of silver and 10 parts of copper. When the copper exceeds this amount, it is oxidised when the alloy is exposed to the air, whence the tarnished appearance of the silver coinage of Prussia, which contains one-fourth of copper.

In English commerce, the purity or fineness of silver is generally expressed as so many pennyweights (dwts.) better or worse than the standard silver, of which the troy pound contains 11 ozs. 2 dwts. of pure silver and 18 dwts. of copper (commonly called *alloy*). Thus the French coin, which contains 24 dwts. of copper in the troy pound, would be described as *worse* 6 *dwts.*, because it contains that quantity less silver than the English coin. Mexican dollars contain 23½ dwts. of copper in the troy pound, being *worse* 5½ *dwts.* Indian rupees sometimes contain only 12 dwts. of copper in the troy pound; hence they are *better* 6 *dwts.*, that is, they contain 6 dwts. more silver than the English standard.

The specific gravity of English silver coin is 10·3, and since all the alloys used to make counterfeits have tin for their chief constituent, they have usually a lower specific gravity, so that the best method by which to test whether a florin, for example, is good, without injuring it, is to ascertain its specific gravity, by weighing it first in the ordinary way, and afterwards when suspended in water, and dividing its weight in air by the loss of weight in water, when the quotient, for a genuine coin, would be 10·3. Of course, if the coin be new, it will be sufficient to ascertain that it weighs just as much as a good florin, and is of exactly the same diameter and thickness, which are readily measured by cutting a slit in a piece of cardboard through which a new florin will exactly pass.

Standard silver is whitened by being heated until the oxygen of the air has converted a little of the copper at the surface into oxide of copper, which is dissolved off by immersing the metal in weak vitriol (diluted sulphuric acid) or in ammonia, or by boiling it in a solution of cream of tartar and common salt. The film of nearly pure silver which then remains at the surface exhibits a want of lustre and is called *dead* or *frosted silver.* It is brightened by burnishing.

Oxidised silver, as it is erroneously called, is made by immersing articles of silver in a solution obtained by boiling sulphur with potash, when the metal becomes coated with a thin film of sulphuret of silver.

The tarnish which is produced upon the surface of silver when exposed to air is also due to the formation of a coating of sulphuret of silver by the action of sulphuretted hydrogen;

the sulphuret of silver is itself black, but a thin film of it upon the surface of the metal often exhibits the rainbow colours caused by the decomposition of the light reflected through it. Tarnished silver is most readily cleaned with a solution of cyanide of potassium, but this salt is so fatally poisonous that its general use should be discouraged. Ammonia (hartshorn) will also remove the film of sulphuret of silver if assisted by friction.

Plated articles are made of an alloy of copper and brass coated with silver. The brass having been melted with the requisite proportion of copper in a black-lead crucible, is cast into bars 3 inches broad, $1\frac{1}{2}$ inch thick and 18 or 20 inches long. The two faces of the bar are carefully smoothed with a file, and a plate of silver $\frac{1}{25}$th inch thick, and somewhat shorter than the bar, is laid upon each face and tied with iron wire. A little saturated solution of borax is allowed to run in round the edges, in order that this salt may melt and dissolve the oxide off the surface of the brass when heated in the furnace. The bar is now laid upon a coke fire and heated until the surfaces of brass and silver have contracted a firm adhesion, when the compound bar is ready for the rolling mill. After rolling, it is cleaned with diluted sulphuric acid. Sometimes the clean copper surface is washed over with solution of nitrate of silver before applying the plate of silver. A thin film of silver would then be chemically deposited upon the surface of the copper and would both prevent oxidation and favour the adhesion of the silver plate.

The *plated wire* used for making toast-racks, &c., is made of copper coated with silver. The strip of silver is bent round into the form of a hollow cylinder, its edges somewhat overlapping; a red-hot cylinder of copper is thrust into this, and the edges of the silver are joined together by rubbing with a steel burnisher. The copper core is then withdrawn, the tube of silver thoroughly cleaned inside, and slipped upon a bright copper rod so as to fit closely, leaving the ends of the copper somewhat projecting. Grooves are made in these ends, into which the silver is forced down, so as to exclude the air from the copper surface inside. The cylinder is made red hot and well rubbed with a steel burnisher, until the silver thoroughly adheres to the copper, which is then drawn into wire.

Electro-plating is now very generally employed for coating articles of baser metal with a film of silver. This art consists in decomposing a solution containing silver, with the aid of a galvanic battery, in such a manner that the metal may be deposited upon the surface of the article to be plated; German silver (p. 173) is generally employed as the material

FIG. 86.—Process of Electro-plating.

for the latter, the articles being then said to be *electro-plated on white metal.* They must be thoroughly cleaned by boiling them with soda, washing with water, and dipping into very weak *aquafortis* (dilute nitric acid) to take off the film of oxide; they are afterwards washed, scoured with sand, again dipped in the weak acid, and finally rinsed in water.

The articles to be electro-plated are suspended by stout copper wires (Fig. 86) in a vessel of wood or earthenware containing a solution of cyanide of silver in cyanide of potassium, of which every gallon contains an ounce of silver. The suspending wires are connected by stout copper wires with the last zinc plate of a galvanic battery consisting of alternate plates of zinc and copper immersed in diluted sulphuric acid; the last copper plate of this battery is connected with a series of silver plates suspended in the silvering liquid opposite to the articles to be coated. The galvanic influence (or *current*) transmitted from the battery causes the decomposition of the cyanide of silver in the solution, the silver being deposited upon the articles to be plated, and the cyanogen, which was combined with the silver in the liquid, uniting with the silver upon the plates of that metal and forming a fresh quantity of the cyanide of silver equal to that which has been decomposed; this dissolves in the liquid and always maintains it of the same strength. The articles are weighed before and after plating, in order to ascertain the amount of silver which has been deposited upon them; it usually amounts to about an ounce and a half upon a square foot of surface.

In order to secure the perfect adhesion of the film of silver, the objects to be plated are sometimes dipped into a solution of nitrate of mercury until they are covered with a thin coating of that metal, before they are immersed in the silvering-bath. They are then *struck* by placing them in the silvering liquid and connecting them with the zinc of a strong battery for a short time, after which they are brushed with fine sand to show that the coating is perfect, and the process of silvering is then proceeded with.

To preserve the silvering solution of uniform strength, the articles to be plated are sometimes kept in motion by attaching the connecting rods to a frame furnished with wheels which travel along a rail on the edge of a vat, and are moved by clock-work or steam-power.

The deposit of silver is without lustre and requires burnishing. It is dried by immersing the article in boiling distilled water, and allowing it to dry by its own heat when removed. When a lustrous deposit is required, one gallon of the silvering liquid is mixed with six ounces of a liquid called *bisulphide of carbon* and set aside for twenty-four hours. Two ounces of this solution are added to twenty gallons of the silvering liquid, and left for twelve hours before use. The action of the bisulphide of carbon in causing a lustrous deposit has not yet received a satisfactory explanation.

Silvering for merely Ornamental Purposes.—Where a very thin film of silver only is required, as for articles not subjected to much wear, advantage is taken of the great malleability of this metal, in which it is surpassed only by gold, to beat it out into exceedingly thin leaves, which are applied to the surface to be silvered. The silver leaf is manufactured in the same manner as gold leaf, to which the reader may refer. It is applied to non-metallic objects with some adhesive liquid, such as gum or size. In covering metallic objects with silver leaf, they are heated to remove grease, and plunged into weak aquafortis to dissolve off the oxide. The surface is next scoured with wet pumice stone, warmed, and again dipped in weak aquafortis, to roughen it, so that the silver leaf may more readily cling to it. If necessary, the surface is further roughened by hatching with a graving tool. The metal is then carefully heated till a thin film of oxide causes it to assume a bluish tint, the leaves of silver applied in successive layers, and well fixed by a burnisher of steel, the object being heated again before every application of the silver.

The process of *dry silvering* upon copper and brass, which is now seldom followed, consisted in applying to the clean surface an amalgam of silver from which the mercury is afterwards expelled by heat. A pasty amalgam is made by dissolving silver in about six times its weight of mercury. The amalgam is applied with a brush made of brass wire dipped into a solution of nitrate of mercury, which, being decomposed by the copper and zinc of the brass, deposits a coating of mercury upon the brush to which the silver-amalgam then readily adheres. The article is then moderately heated to expel the mercury in vapour, when a dead film of silver is left upon the surface, which is afterwards burnished.

For silvering flat surfaces, such as the scales of barometers, the chloride of silver is employed. To prepare this, a piece of standard silver is dissolved in a glass or earthen vessel, with the aid of heat, in *aquafortis* (diluted nitric acid). If any dark powder remains undissolved, it consists of finely divided gold, which is often found in old silver. The solution, which contains nitrate of silver and nitrate of copper, is mixed with common salt dissolved in water. The chloride of sodium (common salt) decomposes the nitrate of silver, forming nitrate of sodium, and chloride of silver, which separates as a white curdy *precipitate.* The liquid is well stirred, the chloride of silver allowed to settle down, the liquid poured off and replaced by fresh water; after this has been repeated several times, the washed chloride of silver is dried in an oven, and finely powdered. One part of the powder is mixed with three parts of pearlash, one part of chalk, and one-and-a-half of salt. The surface of the copper or brass to be silvered is rubbed with a wet cork or leather dipped in this mixture, when the metal decomposes the chloride of silver, forming chloride of copper, and, in the case of brass, chloride of zinc, and metallic silver is deposited.

A mixture of the chloride of silver with ten parts of cream of tartar is said to answer the same purpose.

Silvering on glass is effected by precipitating silver from a solution, in contact with the glass, by certain chemical agents, and being a purely chemical process, will not be considered here. Looking-glasses are silvered with an amalgam of tin (see *Mercury*).

GOLD.

There is no positive evidence that gold exists in Nature in any other than the metallic state, though it is believed by some to exist as a sulphuret in some varieties of pyrites. There are few regions in which small quantities of this metal cannot be discovered, though in the great majority of cases its quantity is too small to pay for the labour of separating it from the other matters with which it is associated.

In England, small quantities of gold are found in the Cornish alluvial deposits which furnish the stream tin ore. In Wales, it has been found near Dolgelly. Ireland has furnished gold from Wicklow, where it is found scantily distributed through sands in the form of gold-dust, and very rarely in small rounded fragments or *nuggets.* In Scotland, the precious metal has been traced in Perthshire, and, very recently, considerable quantities of it have been extracted in Sutherlandshire.

On the continent of Europe, the gold mines of Hungary and Transylvania are the most important. At Königsberg, the metallic gold is disseminated through sulphuret of silver.

In Sweden, gold is found associated with pyrites at Edelfors in Smoland.

The sands of the Rhine contain minute quantities of gold for which they are sometimes washed when work is scarce, although about eight million parts of sand must be washed for one part of gold.

In Spain, the province of Asturias formerly furnished a considerable quantity of gold, but the workings are now neglected.

Italy is by no means destitute of gold. Veins of pyrites containing gold are found in a granitic rock at the foot of Monte Rosa. The sands of some of the rivers on the southern slopes of the Alps also furnish gold.

Siberia yields gold distributed through *hornstone*, a variety of quartz. The sands of Siberian rivers are not considered to be worth washing if they contain less than one part of gold in a million.

The Ural mountains contain some rich gold districts, the metal being found in pyrites, in clay, and in the sands of rivers.

Japan, Ceylon, Borneo and Thibet also contribute to the supply of gold.

Africa seems to have been the oldest and richest of the sources of gold. Sofala, on the coast of Caffraria, has sands abounding in gold-dust, and is reputed to have been the Ophir of the ancients. To the south of the great desert of Sahara, the negroes dig out earth rich in gold-dust to a considerable depth.

In modern times, down to about the year 1850, Brazil, Chili, Peru and Mexico purveyed most of the gold employed throughout the world. Minas Geräes in Brazil was a celebrated auriferous district.

But the discoveries of gold in California and Australia, which are yet fresh in the memory of the present generation, have immensely increased the supplies of the metal. In California, the gold is chiefly *alluvial gold*, being found in the alluvial deposits formed by the Sacramento and other rivers. In Australia, *gold-quartz* is more common, the metal being disseminated in thin plates, and in branch-like fragments, through lumps of quartz-rock. It is also abundant, in the form of gold-dust and nuggets, in the alluvial formation produced by the crumbling down of the rocks containing gold, under the influence of torrents which have carried the gold, together with clay and other matters, into deep gullies, at the bases of the rocks, where the *alluvium* has been deposited upon a bed of pipe-clay, being richest in gold at the lower part of the deposit, on account of the great weight of the metal.

Native gold always contains silver, and generally copper, but in very variable proportions, though the gold from the same district has commonly the same composition. Australian gold is remarkably pure.

The simplest method by which gold is extracted is that of washing the alluvial deposits, the sands of rivers, &c. There are various modes of effecting this, according to the resources of the gold-washers, but in all cases the separation of the gold from the earthy matters depends upon the high specific gravity of the gold (19·3), the lighter earthy matters being carried off by the water.

In Africa, the deposits containing gold are washed by the negroes in the shells of gourds, being well stirred up with water, which is poured off with the earthy matters suspended in it, leaving the gold dust in minute flattened grains at the bottom. The gold is kept in tubes made from the quills of the ostrich or vulture.

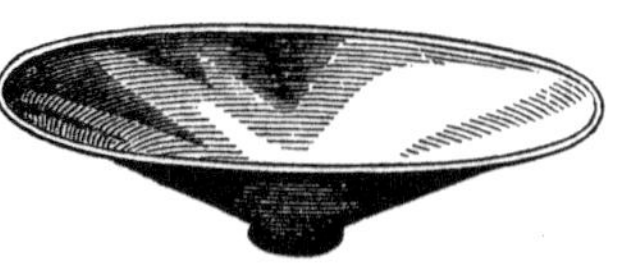

Fig. 87.—Gold-washing Pan.

In South America, the washing is conducted in shallow iron or zinc pans (Fig. 87).

Sometimes the deposits are thrown upon the top of a sloping plank with shallow grooves cut across it, when the grains of gold settle down into the grooves, and the earthy matters are carried on by the stream.

Long shallow troughs of wood are employed by some gold-washers, lined at the bottom with coarse baize, or with tanned skins with the hair upwards. The grains of gold become entangled in these, and when the earthy matters have been washed away, the linings are well beaten over a tank of water to remove the gold.

At the Californian and Australian gold-diggings, a *cradle* (Fig. 88) has been extensively employed. This is a wooden trough about six feet long, resting upon rockers. At the head of it is a grating upon which the alluvial deposit to be washed is thrown. This end of the cradle is about four inches higher than the other, so that a stream of water entering it flows through and escapes at the lower end, left open for this purpose, carrying the earthy matters with it, leaving the particles of gold, with a small quantity of earthy matter, in the trough. These are swept out into a pan, dried in the sun, and freed from the lighter matters by blowing upon them.

In the gold washings of the Ural Mountains, the sands are thrown into boxes the bottoms of which are made of

FIG. 88.—Cradle for Gold-washing.

perforated iron plates. These boxes are placed under a fall of water, in which the sands are stirred up with a shovel. The fine sand and gold dust are washed through on to sloping boards covered with baize, a workman being engaged in sweeping the deposit up the inclined plane with a heather broom. After a second washing on a smaller inclined table, the particles of magnetic iron ore are extracted by a magnet, and the gold dust is melted in a plumbago crucible, when the earthy matters remain upon the surface of the melted gold.

At some of the Russian works, the sifting of the sands is effected in cylinders of perforated sheet iron, which are placed in a sloping position above the washing tables, and made to rotate upon an axis. A stream of water being let in at the upper end, carries the sand and gold dust through the perforations on to the sloping tables, while the large pebbles pass out at the lower end of the cylinder.

The boxes in which the more tenacious alluvial deposits are mixed with water are sometimes provided with agitators worked by horse or steam power, and having knives attached for breaking up and mixing the deposits.

When the gold is disseminated through quartz or some similar rock, this must be crushed in order to extract the gold, an operation attended with great expense, on account of the hardness of the rock. Where it is possible, the rock is rendered more brittle by being heated to redness and quenched with water. The crushing is effected either by passing the gold-quartz between chilled cast-iron rollers, or by means of stampers similar to those employed in the Cornish tin-works.

From the stamped ores, as well as from the auriferous sands which have been concentrated by washing, the gold is sometimes extracted by a process of *amalgamation* similar to that employed in the case of silver.

As practised by the Mexican gold-washers, the process of amalgamation consists in shaking the damp gold-dust, still mixed with foreign matters, with metallic mercury which dissolves the gold. The impurities having been washed away, the amalgam is squeezed in a cloth, when about half the mercury flows out, and the solid amalgam remaining in the cloth is placed in a small iron dish, covered up with green leaves and set over a charcoal fire; a good deal of the mercury vapour is condensed in the leaves, which are renewed from time to time as they get dry.

In the Tyrol, particles of gold exist disseminated through iron pyrites, from which they are extracted by amalgamation. The amalgamating mill (Fig. 89) is a large cast-iron dish (*e*) firmly fixed upon a wooden table. In this dish there is a heavy cone of hard wood (*m*) of the same shape as the dish, and just large enough to leave an interval of half-an-inch between them; several projecting iron ribs are fixed to the under side of this cone, which nearly touch the bottom of the pan, and the upper surface of the cone is hollowed out so as to form a shallow funnel. The wooden cone is connected with an axle (*a*), so that it may be made to revolve at the rate of about twenty turns a minute. About 50 lbs. of mercury having been poured into the iron pan, the auriferous pyrites, previously stamped or ground to a fine powder, is brought into the mill by a stream of water from the spout (G), when it is thoroughly stirred with the mercury by the projecting ridges at the bottom of the wooden cone. In order that no gold may escape being dissolved by the mercury, the pyrites which has been treated in one mill

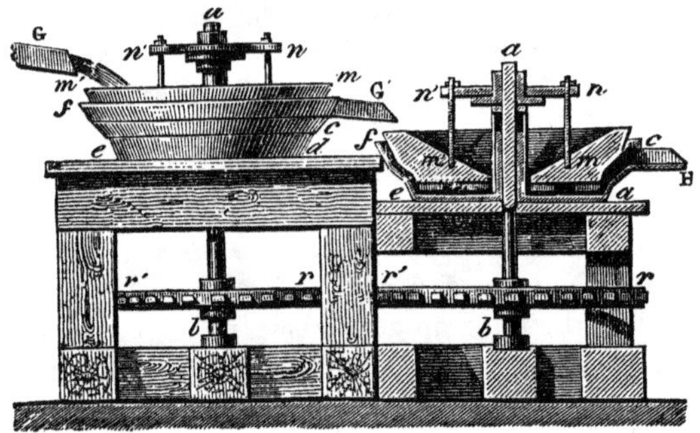

Fig. 89.—Mill for amalgamating Pyrites containing Gold. *r r'*, Toothed wheels for transmitting motion to the axle *a b*.

flows out through a spout (G) into the next, and so on through an entire series. After about a month, the mercury is drawn off and squeezed through wash-leather, which allows the liquid portion to pass through, and retains a soft solid amalgam containing about one-third of its weight of gold, from which the mercury is separated by distillation in the apparatus represented in Fig. 90.

For the extraction of gold from gold-quartz, lead has been employed with great advantage, since this metal, when melted, will dissolve gold just as mercury will at the ordinary temperature. The crushed quartz is fluxed by an addition of lime and clay (see *Iron*), with which is added either metallic lead or galena (sulphuret of lead), or even rich lead slags, with some coal or charcoal to reduce the lead to the metallic state.

The lead containing gold (and silver) is then subjected to the process of cupellation (p. 208).

In Hungary, gold is extracted from iron pyrites associated with quartz, by taking advantage of the property of dissolv-

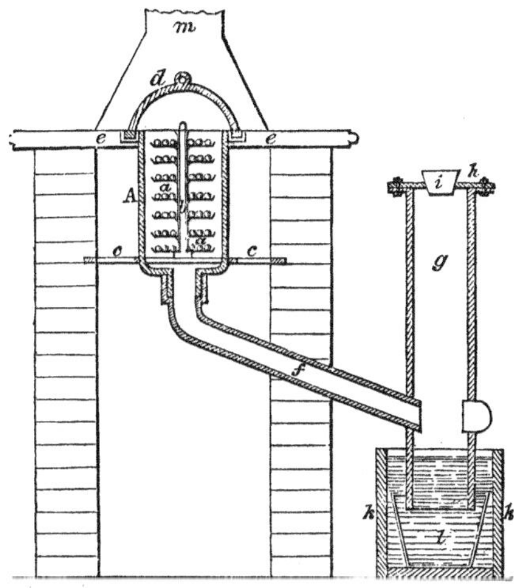

FIG. 90.—Apparatus for distilling the Amalgam of Gold. *a*, Dishes for receiving the amalgam, attached to the pillar *b*. A, Iron cylinder heated by a fire on the grate *c*. *d*, Iron dome. *e*, Iron cover of the furnace. *f*, Tube for vapour of mercury. *g*, Cylinder in which the mercury condenses, closed by an iron plate *h* and a wooden plug *i*. *k*, Water-tank. *l*, Vessel for collecting the mercury. *m*, Chimney.

ing gold possessed by sulphuret of iron. The pyrites is roasted in heaps with brushwood, to convert a part of the bisulphuret of iron into oxide of iron, and a part into sulphuret of iron. The roasted ore is fused with an addition of lime, when a slag is formed by the combination of the silica (quartz), the oxide of iron, and the lime, whilst the sulphuret of iron fuses, dissolves the gold, and forms a matt beneath the slag.

This sulphuret of iron is roasted so as to convert it into

oxide, and fused with a fresh quantity of the auriferous pyrites and the requisite proportion of lime, when a fresh quantity of the matt will be obtained, containing the gold from the two charges of pyrites. This operation is repeated until a sufficient quantity of gold has accumulated in the matt of sulphuret of iron, which is then melted down with lead; this metal extracts the gold from the sulphuret of iron, and the latter remains in a melted state upon the surface of the lead. The lead is afterwards cupelled in order to extract the gold, and since lead containing silver is generally employed, this metal is left on the cupel alloyed with the gold.

Whenever ores of copper, lead, or silver contain gold, the latter is always present in the metal extracted from them, and is recovered from those metals by the processes described in the article on Silver.

In order to separate the gold from the silver with which it is commonly alloyed, whether it has been obtained by washing, or by any of the above processes of extraction, the alloy of silver and gold is heated with sulphuric acid, which converts the silver into sulphate of silver, capable of being dissolved by water, and leaves the gold untouched.

Parting by Sulphuric Acid.—The alloy of silver and gold, in which, of course, the former metal always predominates, is melted either in wrought-iron or plumbago crucibles, and poured into water in order to *granulate* it or divide it into a flaky condition, exposing a large surface to the action of the acid. The granulated metal is dried, weighed, and boiled with oil of vitriol (concentrated sulphuric acid). When the alloy is rich in gold, the operation is performed in platinum alembics or stills, but in the more common case, where the silver contains only a few grains of gold in the pound, cast-iron pans are employed; each pan (about two feet wide) has an iron lid, from which a bent pipe passes down into an air-tight leaden tank, where the vapour of sulphuric acid which escapes during the boiling may be condensed. A large quantity of sulphurous acid gas passes off during the operation, and this is conducted, from the leaden tank, by a pipe of the same metal, into a large leaden chamber, 30 feet long by 10 feet wide, and 6 feet high, in which it is reconverted, by an appropriate chemical process, into oil of vitriol, which is used over again.

One fire is made to heat two of the iron pots, in which the granulated silver is placed, together with twice its weight of concentrated sulphuric acid, which is gently boiled until the silver is entirely converted into sulphate of silver, forming a pasty mass consisting of minute crystals. This is taken out by cast-iron ladles, and thrown into leaden cisterns, where it is stirred up with water, and boiled by passing steam into it through perforated leaden pipes connected with a boiler. The boiling water dissolves the sulphate of silver, and the finely-divided gold is left as a black powder, which, when accumulated in sufficient quantity, is well washed and dried. It still retains a small proportion, varying from $\frac{1}{30}$th to $\frac{1}{60}$th of its weight, of silver.

The solution of sulphate of silver is drawn off, by leaden siphons, into leaden troughs, where it is left in contact with shavings of copper. This metal enters into solution, forming sulphate of copper, and separating the silver in a finely-divided state, as a grey powder; this is allowed to settle down, the solution of sulphate of copper run off into another cistern, and the silver washed with fresh water, drained, and compressed by hydraulic pressure, in a square cast-iron box, which makes it into cakes of 60 lbs. each. These are dried, melted in plumbago crucibles, and cast into ingots. Cast-iron crucibles strengthened by shrinking hot iron hoops upon the cold crucibles are sometimes employed, but since they become impregnated with silver, the latter must be extracted by melting some lead in them when they are worn out.

The solution of sulphate of copper formed in displacing the silver by copper is evaporated in shallow leaden pans, to a proper strength, and the sulphate of copper allowed to crystallise out on cooling. The liquid remaining after the last crystals have separated contains the excess of sulphuric acid which has been employed in the process, very little sulphate of copper being left in it, because this salt is almost insoluble in moderately strong sulphuric acid. This liquor is boiled down in a platinum still, until the water has boiled away, and the concentrated sulphuric acid is left in the still, ready to be employed for the treatment of a fresh quantity of silver.

The sulphate of copper (blue vitriol) obtained in this process is a salt for which there is a considerable demand; it is largely used for dressing grain intended for seed, to prevent smut. It is also employed in dyeing and calico-printing, and in many other branches of industry, as well as in several forms of galvanic battery.

When there is no market for the sulphate of copper, the solution of the salt is decomposed by scrap iron, as in the case of the blue water at Anglesea (p. 128), to recover the metallic copper.

In such works for the refining of gold and silver, the processes can be conducted economically only when great care is taken to avoid the loss of any particles of the precious metals. Thus all the old crucibles are ground and treated with mercury in the amalgamation mill, and after as much gold and silver as possible have been thus extracted, the residues are sold to the *sweep-washers*, who extract a little more by melting with lead. The very dust off the floors is collected and treated in a similar manner. One part of gold can be profitably extracted from 2,000 parts of alloy by this process of parting by sulphuric acid. Its introduction has affected the metallurgy of gold in the same way as Pattinson's process did that of silver, much old silver plate having been treated by it for the sake of the gold which had not been found worth extracting by the older and more expensive method of parting by nitric acid.

When the alloy contains copper as well as silver and gold, it may also be treated in the same way, the copper being removed, with the silver, as a soluble sulphate; but the process does not succeed well with an alloy containing more than 75 parts of copper in 1,000, so that, if it be richer, it is either melted with more silver, or is cupelled with lead (p. 208), in order to reduce the copper to the right proportion.

Nor should the alloy contain more than one-fifth of its weight of gold, or the sulphuric acid will not extract the silver. When platinum stills are employed for parting by sulphuric acid, it is necessary that the alloy should be free from lead and tin, which are apt to melt upon the bottom of the still and seriously to corrode the platinum.

Parting of Gold and Silver by Nitric Acid.—Silver is easily dissolved by nitric acid and converted into nitrate of silver, but this acid, if pure, does not attack gold. If the nitric acid contains chlorine, however, it will dissolve some of the gold, so that it is always necessary to test it by adding a little solution of nitrate of silver, which will render it milky, from the separation of the insoluble chloride of silver, if any chlorine be present. An alloy containing more than one part of gold to three parts of silver is very little affected by nitric acid, so that it becomes necessary to fuse very rich alloys with so much silver that the gold shall form only one-fourth of the alloy; this is the origin of the term *inquartation* or *quartation*, used in speaking of this process.

The alloy, in a granulated state, is heated with twice its weight of moderately strong nitric acid (sp. gr. 1·32) in a still made of platinum, glass, or earthenware, connected with an apparatus for condensing the vapours of nitric acid which pass off. Whilst the silver is being dissolved, a large quantity of red gas is evolved, resulting from the action of the silver upon the nitric acid, and when this is no longer perceived, the silver is known to be dissolved. The still is then cooled, the solution of nitrate of silver drawn off, and the undissolved gold boiled with a little more nitric acid to extract any remaining silver. It is then washed with water, dried, melted, and cast into an ingot.

In order to recover the silver from the nitrate, hydrochloric acid is cautiously added, so as to separate the bulk of the silver as the insoluble chloride, leaving the nitric acid in the solution, which may be used again, if care be taken to leave a little nitrate of silver undecomposed in the solution, so as to ensure the absence of chlorine. The separated chloride of silver is washed with water, moistened with sulphuric acid, and some bars of zinc placed in it, when chloride of zinc is formed and dissolved, the silver being left in the finely-divided metallic state. The rest of the zinc is then taken out, the silver allowed to remain in contact with dilute sulphuric acid to dissolve any particles of zinc, then thoroughly washed with water, dried, melted, and cast into ingots.

Refining of Gold.—The gold obtained by parting with sulphuric acid is refined by mixing it with one-fourth of its weight of dried sulphate of soda, and treating it, in an iron pan, with oil of vitriol, to the amount of three parts for every five parts of sulphate of soda. Heat is applied as long as any vapours of sulphuric acid escape. This is repeated a second time, but without driving off the whole of the sulphuric acid. The mass is then boiled with sulphuric acid, when the gold alone is left, and is melted with a little saltpetre, which extracts a little platinum, before casting it into an ingot. At the Russian mint, the re-melting is effected in a small reverberatory furnace, with a cavity in which the gold collects. The explanation of this process is simply that the sulphuric acid combines with the sulphate of soda, and may then be raised to a higher temperature without vaporising than is possible with uncombined sulphuric acid. The higher temperature employed enables the sulphuric acid to attack the remainder of the silver. (See Addendum, p. xi.)

Extraction of Gold from Gold-quartz in the wet way.—It has been proposed to avoid the expensive process of amalgamation, by digesting the pulverised gold-quartz with $\frac{1}{100}$th part of black oxide of manganese and some muriatic acid, in an earthen vessel, for twelve hours, when chlorine is generated, which dissolves the gold in the form of chloride of gold, from which the metal may be separated in a finely-divided state by adding to the liquid a solution of copperas (sulphate of iron). The dark powder of gold thus separated is washed with water, dried, and melted down. When the gold contains much silver, common salt is employed to dissolve the chloride of silver, which would otherwise protect the gold from the action of the chlorine.

Plattner's Process, which was employed with economy for the extraction of gold from the abandoned residues of roasted pyrites at Reichenstein, in Upper Silesia, containing less than 1 oz. of gold per ton, is rather a chemical than a metallurgical process, and consists in treating the fine powder, in a moist state, with chlorine gas, which converts the gold into a soluble chloride ; this is washed out with water, and treated with sulphuretted hydrogen, which separates the gold as an insoluble black sulphuret, leaving the iron, &c., in the solution. The sulphuret of gold is heated to expel a part of the sulphur, dissolved in a mixture of hydrochloric and nitric acids, and separated from the solution in the pure metallic state by sulphate of iron. The precipitated gold is washed, and melted down with a little borax and saltpetre.

Gold dust and nuggets of gold never consist of the pure metal, but always contain silver, and sometimes copper and small quantities of other metals, such as antimony and bismuth. Grains of platinum and its allied metals are also very commonly found in alluvial gold. The purest native gold has been found at Giron, in New Grenada, containing only $\frac{1}{100}$th part of silver. Some Californian gold contains as much as nine parts of silver and nearly one part of copper in a hundred parts. Californian gold also sometimes contains small grains of an extremely hard alloy of osmium and iridium, which occasion great injury to the die in coining, since they remain unchanged when the gold is cast into ingots.

At the American mint, the Californian gold, which contains about $\frac{1}{1000}$th of its weight of the osm-iridium alloy, is melted with thrice its weight of silver, which lowers its specific gravity, and allows the osm-iridium to settle to the bottom. The greater part of the melted metal is ladled out, leaving the rest very rich in osm-iridium at the bottom. This is repeatedly melted with silver, by which the proportion of gold is still further diminished, and ultimately the mixture of osm-iridium with silver and a little gold is boiled with sulphuric acid, which extracts the silver, leaving the osm-iridium mixed with some powdered gold, which may be removed by washing.

When gold contains platinum or palladium, it is very much lighter in colour. These metals cannot be separated by the ordinary refining processes. The presence of lead or antimony, even in very minute proportion, is found to render gold extremely brittle.

Australian gold is frequently brittle from the presence of lead or antimony. It is sometimes refined by stirring a little corrosive sublimate (chloride of mercury) into the melted gold, when the chlorine combines with the base metals, forming chlorides which are expelled in the form of vapour, together with the liberated mercury. F. B. Miller has introduced an improved process for refining such gold by forcing into the melted metal a current of chlorine gas, through the stem of a tobacco-pipe. The chlorine converts the silver present in the gold into chloride of silver, which collects, in a melted state, upon the surface of the gold, whilst any arsenic, antimony, bismuth, lead, or zinc, is also converted into chloride, and driven off in the form of vapour. The silver is afterwards easily extracted from its chloride.

Perfectly pure or *fine gold* is nearly as soft as lead, far too soft therefore to resist the wear to which it would be subjected in coinage and gold plate. The alloy used for coin, in England, consists of 11 parts of gold and 1 part of copper, which is harder and more fusible than pure gold. Formerly, the gold was alloyed with silver, or with silver and copper, and this latter alloy is still employed by goldsmiths. The guinea was composed of 11 parts of gold, ½ part of silver, and ½ part of copper. The specific gravity of sovereign gold is 17·157 (that of pure gold being 19·3). The safest method of ascertaining whether a sovereign is genuine consists, as in the case of silver coin (p. 233), in showing that it has the same size and weight as a sovereign known to be genuine. A new sovereign weighs 123¼ grs., but it is a legal tender as long as it is not less than 122½ grs. in weight. So hard is the alloy, that with proper wear, a sovereign will circulate for eighteen years without falling below the legal standard.

The gold coin of the United States and of France contains only 9 parts of gold to 1 part of copper.

The fineness or purity of gold is commonly expressed by stating how many carats of gold are present in 24 carats of the alloy. Thus pure gold would be 24 carats fine; sovereign gold, 22 carats fine.

Fractions of a carat are expressed in grains (4 grains are equal to 1 carat) and eighths of a grain; thus French gold coin would be styled of 21 carats $2\frac{3}{8}$ grs. fine, or *worse O carat* $1\frac{5}{8}$ grs., implying that it contained so much less gold than the English standard.

The fineness may of course be judged of, as in the case of sovereign gold, from the specific gravity, since the specific gravities of silver and copper (respectively, 10·5 and 8·9) are so much lower than that of gold (19·3). But this test has been found fallacious in a case where bars of platinum (sp. gr. 21·5) were coated with gold and sold as solid ingots of that metal, which is more than twice the price of platinum. Gold of 18 carats fine has the specific gravity 16·8.

The goldsmith or pawnbroker generally tests the gold by touching its surface with a stopper wetted with *aquafortis* (nitric acid), which produces a green stain upon the metal when a very large proportion of copper is present, nitrate of copper being then produced. This is, at best, a rough test, and would of course fail altogether if the surface only of the base alloy were coated with fine gold.

The use of the *touchstone* admits, in practised hands, of a far more exact determination of the value of the alloy, and unless it be pretty thickly coated with richer metal, deception is more easily detected. The touchstone is a piece of black basalt, or even of black slate, over which the gold to be tested is drawn so as to leave a streak of fine particles of the metal upon the surface; this streak of course remains untouched when moistened with nitric acid, but if a streak of any base alloy (of copper and zinc for example), made to imitate gold, be made upon the surface of the touchstone, the nitric acid will immediately dissolve it.

The acid employed in this test is generally mixed with a minute proportion of hydrochloric acid (98 parts by weight of nitric acid, of sp. gr. 1·34, 2 parts of hydrochloric acid, of sp. gr. 1·173, and 25 parts of water). The streak is not apparently affected by the acid if the gold is not below 18 carats fine; by making several streaks in succession, or by grinding off a part of the surface upon the touchstone, any error arising from a thin external coating of fine gold may be avoided; the feather of a pen, or a glass rod, serves for moistening the streaks with the acid.

In order to determine by the touchstone the proportion of gold which is present in the alloy, the streak is compared with that made by a series of *touchneedles* composed of alloys containing gradually diminishing quantities of gold. In experienced hands, the quantity of gold may thus be ascertained with an error of not more than one part in a hundred.

Fig. 91.—Cupel.

The exact assay of the alloys of gold is an imitation, on the small scale, of the metallurgic processes of inquartation, cupellation and parting. A weighed quantity of the alloy, say 6 grs., is wrapped in a piece of thin paper together with three times its weight of pure silver, and added to twelve times its weight of pure lead already melted in a bone-ash cupel

(Fig. 91), heated in a muffle (Fig. 92) or arched clay oven, with slits for admitting air, which is placed in a brisk fire (Fig. 93). The lead and copper are both converted into oxides, the oxide of copper dissolving in the oxide of lead, and both being absorbed by the bone-ash. After the *brightening* (p. 211) the alloy of silver and gold which is left on the cupel is hammered flat, annealed by heating to redness, rolled out thin, coiled up into the form of a tube, and boiled, first with weak nitric acid (sp. gr. 1·18), and then with a stronger acid (sp. gr. 1·28) in order to extract the silver; the gold is left in the form of a little tube (*cornette*) having much the appearance of red earthenware. It is well washed with water, carefully transferred to a small crucible without breaking it, dried, and heated to redness in the muffle, when it shrinks, and assumes the ordinary appearance of gold. Its weight is ascertained by a very accurate balance, and multiplied by four (six grains having been assayed) to express the fineness of the gold. To avoid errors, the exact assayer commonly passes a *proof* or weighed quantity of pure gold through the same process, at the same time, with the addition of a proportion of copper about equal to that in the alloy, and corrects his assays by deducting from the weight of the gold finally obtained the increase which the pure gold was found to have experienced in consequence of its having retained traces of lead, silver and copper.

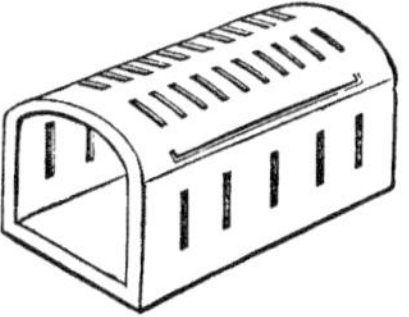

FIG. 92.—Muffle.

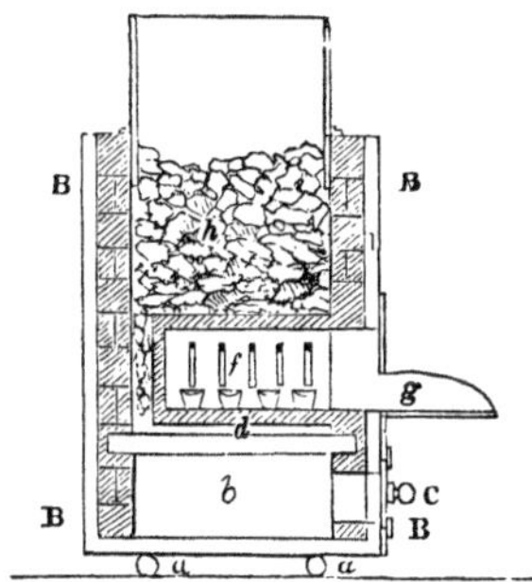

FIG. 93.—Assay by Cupellation. *a*, Iron castors on which the furnace moves. *b*, Ash-pit. c, Damper. *d*, Grate. *f*, Muffle containing the cupels. *g*, Mouth-plate. *h*, Fuel.

The alloy of copper and gold is much redder than pure gold; the addition of silver whitens it, and its surface may be brought to any shade of gold colour by heating it until a portion of the copper is oxidised, and dissolving out this oxide with an acid. Goldsmiths commonly *colour* their gold by boiling it for twenty minutes with a mixture of 1 part of common salt, 2 parts of saltpetre, 1 part of alum and 4 parts of water, the mutual action of which would result in the production of a little hydrochloric acid (from the chloride of sodium, common salt), and a little nitric acid (from the saltpetre, nitrate of potash); this *pickle* dissolves not only the copper, but some of the gold, which is recovered by precipitating it with solution of copperas (p. 251).

The great value of gold, and its perfect freedom from alteration by exposure to the atmosphere, have led to many devices for making the smallest quantity of the precious metal cover the greatest extent of surface, by taking advantage of its extreme malleability and ductility, in which it far surpasses all other metals.

Gold-beating.—Pure gold may be extended by hammering so as to present a surface 650,000 times as large as it originally possessed, but for this purpose it must be perfectly pure or fine gold, a very small quantity of any other metal materially injuring its malleability. Gold alloyed with copper or silver is, however, often beaten into moderately thin leaves, when a colour different from that of fine gold is required, but it is of course much more liable to tarnish. Dentists' leaf gold for stopping teeth is generally beaten from fine gold, because it is more easily pressed into a compact form.

The gold is melted in a crucible with a little borax, which prevents it from sticking to the crucible, and poured into a small cast-iron mould, warmed and slightly oiled, in which it forms an ingot of ¾ inch in width, and 2 ozs. in weight. This ingot is annealed by burying it in hot ashes, and hammered upon a steel anvil until its thickness is reduced to

$\frac{1}{8}$th inch, the metal being once or twice re-heated or annealed during the process. The gold is then passed between perfectly cylindrical steel rollers, which are turned at exactly the same rate in opposite directions, until it is reduced to a riband so thin that a square inch of it weighs $6\frac{1}{2}$ grains. It is of course necessary to anneal the gold several times during the rollings.

The riband is now exactly divided, with the help of a pair of compasses, and cut up into pieces of one inch square. One hundred and fifty of these pieces are piled up alternately with pieces of tough paper or vellum (prepared calf-skin) four inches square, rubbed over with a little fine plaster of Paris, to prevent the gold from sticking. Twenty vellums are placed above, and twenty below the pile, which is then firmly secured by passing two strong belts of parchment across it.

The beating is performed with a hammer weighing about 16 lbs., having a circular face four inches in diameter and somewhat convex. The pile is placed upon a heavy block of marble, nine inches square, sunk into a very strong wooden bench. The workman uses the hammer with one hand, directing the blows well in the middle of the pile, and turning it occasionally. After a time, the packet is opened, the middle leaves are shifted to the outside, and the beating is continued until the leaves have been extended to nearly the same size as the vellums. They are then taken out of the pile, and each leaf is cut with a sharp knife into four equal squares, which will measure about an inch each way. These are made into packets, as before, with *gold-beaters' skin*, a membrane which is separated from the outer surface of the intestines of the ox, and is prepared for use by hammering it between folds of paper in order to extract the grease, and steeping it in an infusion of nutmeg and cinnamon, in order to preserve it; it is then dried and rubbed over with plaster of Paris. The packets made up of gold-

S

beaters' skin, with alternate layers of gold, are beaten as before, but with a ten-pound hammer, for about two hours, the packet being skilfully rolled and bent now and then, in order to loosen the leaves. When these are again extended to about 4 inches square, they are spread out upon a leathern cushion, and cut into four equal squares by applying a cross of cane cut to sharp edges, and fixed upon a board. These squares are again made up into packets with gold-beaters' skin, and hammered out a third time, with a seven-pound hammer, to about $3\frac{1}{2}$ inches square. They are then dexterously lifted off the skin, with a delicate pair of wooden pincers, spread out upon a leathern cushion by blowing them flat down, and cut down to one size by a square of sharp-edged cane fixed to a board, and pressed upon the leaf of gold extended on the cushion. The gold leaves are then packed between the leaves of books made of printed paper, rubbed over with red chalk to prevent adhesion; there are usually twenty-five leaves in each book, their average thickness being $\frac{1}{282000}$th of an inch. Mechanical power has been lately substituted for manual labour in gold-beating.

When one of these leaves is held up to the light, it exhibits a beautiful green colour, and if it be rendered still thinner, either by beating, or by floating it upon a very weak solution of cyanide of potassium, which slowly dissolves it, it transmits, when taken upon a glass plate and held up to the light, a blue, violet, or red light, in proportion as its thickness diminishes. Even when it is so transparent that one may read through it, the yellow colour and lustre of the gold are still visible by reflected light. These varying colours of finely-divided gold are turned to account in the colouring of glass and in painting on porcelain.

Gold Thread is made by covering a cylinder of silver with gold leaf, and drawing it through a wire-drawing plate until it is reduced to the thinness of a hair. In this manner, one grain of gold is made to cover 364 feet of wire. For making gold lace, this wire is flattened by passing it between rollers, and is twisted by machinery round a thread of yellow silk, which is then made up into lace or braid.

Gilding.—The most obvious method of gilding consists in applying gold leaf to the object required to be covered, an art requiring great delicacy and skill, the description of which does not fall within the scope of this work.

Wash-Gilding is effected with an amalgam of gold, prepared by dissolving one part of fine gold in eight parts of mercury. The gold is laminated, placed in a crucible, heated to faint redness, and thrown into the requisite quantity of mercury, also moderately heated in a crucible, whilst the metals are stirred with an iron rod hooked at the end. When all the gold is dissolved, the amalgam is poured into water, and squeezed in wash-leather to separate the excess of mercury. A pasty amalgam is thus obtained, containing about one-third of its weight of gold.

The metals generally employed for wash-gilding are brass, and copper alloyed with one-seventh of brass, nickel being sometimes added. The article to be gilded, after being heated to redness in a charcoal or peat fire and cooled slowly, is dipped in very dilute sulphuric acid, rubbed with a hard brush, washed and dried. It is then dipped in pretty strong nitric acid (sp. gr. 1·33), well washed, and dried by rubbing with bran. By this treatment, all the oxide has been removed from the surface, which has been also sufficiently corroded or roughened by the acid to favour the adhesion of the amalgam of gold. This is applied to the surface with a brush made of fine brass wire, which is dipped into a solution of nitrate of mercury made by dissolving 100 parts by weight of mercury in 110 parts of nitric acid (sp. gr. 1·33) and diluting the solution with 25 times its weight of water. The brush being wetted in this solution, which coats it with a thin film of mercury (p. 237), is rubbed upon a lump of the amalgam, which is then brushed over the article to be gilded; this is next held over glowing charcoal, and well turned about, until it is hot enough to make a drop of water hiss, when the whole of the mercury is expelled in the form of vapour, leaving a film of dead gold upon the surface, which is then rubbed with a wet burnisher of hæmatite. Several curious processes, very difficult of explanation, are employed by persons experienced in the art of gilding, for giving the required shades of colour and degrees of lustre to the film of gold left by the mercury. One of these, intended to produce red gold, consists in coating it with *gilders' wax*, containing wax, red ochre, verdigris (acetate of copper) and alum. The object, having been coated with this composition, is strongly heated in the flame of a wood fire, quenched in water, and scrubbed with vinegar. This process probably reddens the gold by alloying it with a little copper from the verdigris.

Buttons, trinkets, &c., made of brass or copper, are coated with gold by immersing them in a boiling alkaline solution containing that metal. 2,400 grains of fine gold are dissolved in a mixture of 21 (avoirdupois) ozs. of nitric acid (sp. gr. 1·45), 17 ozs. of hydrochloric (muriatic) acid (sp. gr. 1·15), and 14 ozs. of distilled water. This solution of (chloride of) gold is mixed with 4 gallons of distilled water, and 20 lbs. of bicarbonate of potash, and boiled for two hours. The articles having been thoroughly cleaned from oxide, are suspended from a hoop of brass or copper-wire, and moved about in the boiling liquid until they are judged to have acquired a sufficient coating of gold, which is the case after less than a minute with small articles. The gold is deposited, in this process, in consequence of the copper taking its place in chemical combination in the liquid; the film is bright and not dead as in the amalgamation process, but it may be deadened, if required, by dipping the articles in solution of nitrate of mercury (p. 259) before exposure to the gilding solution, when they will become coated with mercury, which will form an amalgam with the gold deposited from the solution, and this will be left as a dead film on expelling the mercury by heat.

Electro-gilding is effected, like the corresponding process of silvering (p. 235), by immersing the thoroughly-cleansed articles to be gilded in an appropriate solution of gold, and connecting them by metallic wires with the zinc end of a galvanic battery of which the copper end is in metallic connection with a plate of gold, which is gradually dissolved, replacing the gold which has been deposited. The gilding bath is heated by steam to about 200° F., which facilitates the deposition of the metal. A solution for electro-gilding may be prepared by dissolving 1,550 grains of gold in a mixture of 14 (avoirdupois) ozs. of nitric acid (sp. gr. 1·45), 11 ozs. of hydrochloric acid (sp. gr. 1·15), and 10 ozs. of water. To this liquid, containing the chloride of gold, a solution of cyanide of potassium is added carefully, as long as any cyanide of gold is separated. This precipitate is allowed to settle, the clear liquid drawn off, and the precipitate redissolved in some more solution of cyanide of potassium. Enough water is then added to bring the solution up to five gallons.

In gilding polished iron and steel, they are heated until they assume a blue tint, and successive coatings of gold leaf are applied with a burnisher, a gentle heat being employed after every coating except the last, which is burnished cold.

MERCURY.

This metal, also called *quicksilver*, is of very rare occurrence in Great Britain, the whole of the mercury employed in the arts being imported from Austria, Spain, California, and China, the largest quantity of the metal being now furnished by California. A considerable quantity is found in the metallic state, either collected in cavities, or disseminated in globules throughout its ore *cinnabar*,* which has the same composition as *vermilion*,† being a compound of mercury with sulphur, containing, when pure, 86 parts of mercury in the hundred. Native mercury sometimes contains silver. Cinnabar is usually met with in moderately hard dark brown masses which are very heavy (sp. gr. 8·2), and exhibit a red colour when scraped with a knife. Some specimens have a red colour without scraping, and all yield a powder of a more or less bright red colour.

Native *calomel*, a combination of mercury with chlorine, is sometimes found in very small quantity associated with cinnabar.

A rich deposit of cinnabar is reported to have been lately discovered in Borneo.

The extraction of mercury from its ore is effected, like that of zinc, by distillation, but since this metal is converted into vapour at a much lower temperature than zinc (mercury boils at 662° F.), it requires much more elaborate arrangements for the condensation of its vapour into the liquid form, and in many works a considerable quantity of the metal is wasted, on account of the imperfect character of the condensers, the escaping vapour proving most injurious to the health of the persons employed. Indeed the metallurgy of mercury, in all its branches, is lamentably injurious to health, the miners as well as the smelters being liable to salivation, and seldom living to an old age.

The mines at Almaden, a town of La Mancha, in Spain, have been the most productive, the cinnabar being found in large veins surrounded by sandstone and clay slate. The mercury is extracted by simply roasting the ore, when the oxygen of the air converts the sulphur into sulphurous acid gas, and the mercury passes off in vapour.

* Derived immediately from the Latin *cinnabaris*, applied not only to this ore, but apparently also to red-lead, and to the red gum-resin known as *dragon's blood*, from the Indian name of which, *cinoper*, the word appears to have originated.

† Probably from *vermeil*, French for red coral.

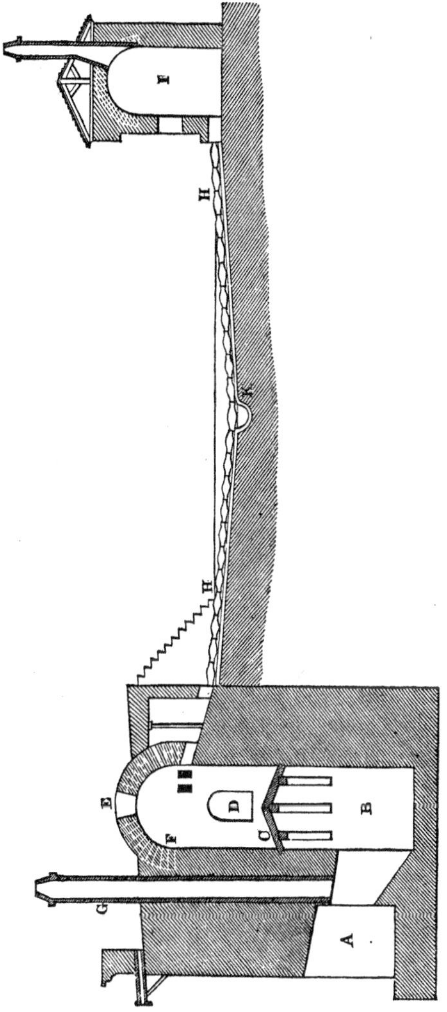

Fig. 94.—Aludel Furnace for the extraction of Mercury. A, Opening for charging the fuel. B, Fire-space. C, Arch for sustaining the ore. D E, Openings for charging the ore. F, Chamber for receiving the ore. G, Chimney. H H, Rows of aludels. I, Condensing chamber. K, Gutter for mercury.

The roasting furnace consists of a fire-place in which a wood fire is maintained, the flame of which kindles the ore stacked upon an arch of fire-brick (C, Fig. 94), provided with openings for the passage of the flame. Upon this arch some large blocks of sandstone, very poor in cinnabar, are piled; above these, small fragments of richer ores, over which again are placed the cakes made up by kneading with clay the finely-divided mercury which has condensed from the fumes without running together into liquid metal, as well as any cinnabar which has escaped in vapour unchanged during a previous roasting, and has been condensed again as a black

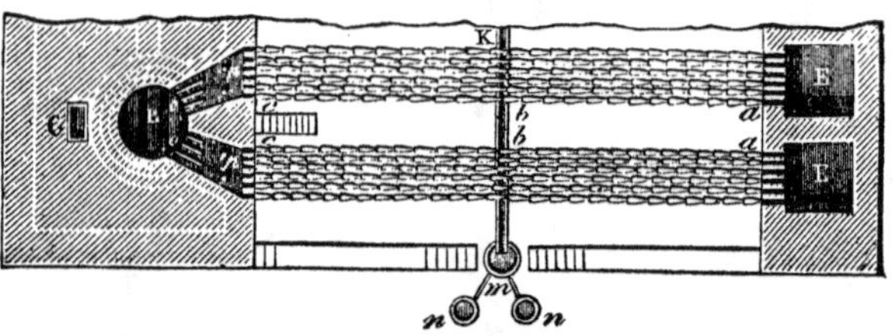

FIG. 95.—Plan of Aludel Furnace for extraction of Mercury.

powder. The air, which enters through the openings in the brick arch, furnishes oxygen to the sulphur of the sulphuret of mercury, and converts it into sulphurous acid gas, while the mercury does not combine with oxygen, but separates in the metallic state, being converted into vapour, partly by the heat of the fire beneath, partly by that produced in the combustion of the sulphur. The sulphurous acid gas and the vapour of mercury pass out through flues (*o*, Fig. 95) in the side of the furnace, into the condensing apparatus, which consists of 300 pear-shaped stoneware pipes, called *aludels* (Fig. 96), open at both ends, and fitted into each other, the joints being cemented together with clay. These are arranged so as to form twelve separate flues, each com-

FIG. 96.—Aludels.

posed of 25 aludels, The first 12 of each set, nearest to the furnace, incline downwards, towards a central gutter (K), whilst the last twelve ascend a corresponding inclined plane, the centre aludel of the series being perforated in the lower side, so that the condensed mercury may run out into a gutter which conducts it into receiving basins (*m*, *n*, Fig. 95) underneath the furnace. The vapour of mercury which escapes condensation in the aludels passes, together with the sulphurous acid gas, into a chamber (I, Fig. 94) in which a further considerable quantity of the mercury is condensed to the liquid state. The sulphurous acid gas eventually escapes through the chimney.

The imperfect character of this condensing apparatus is evident; the loss of mercury vapour through the leakage of the numerous joints, and through the openings in the lowest aludels, must be very great, so that it is not surprising that only 10 parts of mercury should be extracted from a hundred parts of rich ore. Each roasting lasts about twelve hours, and the furnace requires three or four days to cool before receiving a fresh charge, affording a striking contrast to the system of nearly continuous working adopted in the metallurgic processes of this country.

The mercury mines at Idria in Austria, which are now probably more important than those at Almaden, were formerly worked by state prisoners and criminals, on account of the unhealthy nature of the occupation. In these mines the ore is found both in limestone and in a bituminous slate, and contains a considerable proportion of bituminous matter. Previously to the year 1794, the aludel furnace just described was also employed at Idria, but it was then superseded by a very large brick structure containing a series of condensing chambers. The roasting furnace (Fig. 97) consists of a grate upon which wood is burnt, surmounted by three perforated arches (*n p r*) of fire-brick, for receiving the ore, the large fragments being placed upon the lowest arch, the smaller pieces on the next, whilst the uppermost supports a number of shallow earthen dishes (*s*) containing the dust of the ore and the mercurial soot from former operations. Air is admitted to the heated ore through small passages in the walls,

Fig. 97.—Extraction of Mercury at Idria. G H, Passages from which air enters through flues into the space occupied by the ore. C, Condensing chambers.

and the sulphurous acid and vapour of mercury are drawn, by the two chimneys, through six condensing chambers (C D, Fig. 98) on each side of the furnace, so connected by narrow openings that the vapours are forced to pervade one

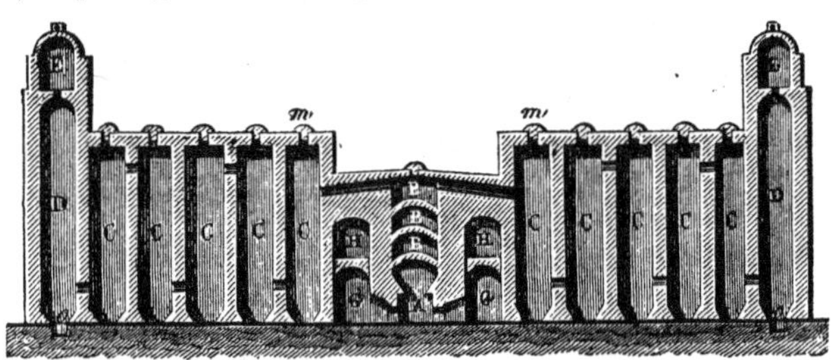

Fig. 98.—Section of Idrian Furnace and Condensing Chambers. A, Fireplace. B, Arches upon which the ore is placed. C D E, Condensing chambers. G H, Air channels.

chamber before entering another. In the last chamber (D) of each series, which is surmounted by the chimney, the last portions of mercury are condensed by a cascade of water. The greater portion of the mercury is condensed into the liquid form in the first three chambers of each series, and is collected in an underground gutter which conveys it into a tank, from which it is ladled out to be filtered through cloth, and put into the wrought-iron bottles, containing about 60 lbs. each, in which it is imported into this country. The three last of each series of condensing-chambers receive the remainder of the mercury chiefly in the form of dust or soot, consisting of finely-divided mercury, with some sulphuret of mercury which has escaped in vapour, and some carbon from the bituminous matter in the ore. About a week is required to complete the distillation, including the time required for cooling the furnace. Only about 8½ parts of mercury are obtained from 100 parts of ore. In 1803, a fire broke out in the mines at Idria, and the combustion was sustained by the bituminous matter in the ore, so that it became necessary to flood the workings with water. The vapours of mercury evolved proved very injurious to the health of the neighbourhood.

A process similar to that employed at Idria is followed at New Almaden, in California, for the extraction of the mercury.

Within the last few years, a greatly improved apparatus has been employed at Idria. The ore, in fragments as large as a fist, is thrown through a hopper into a deep cylindrical fire-place, upon which a little wood and coal are burnt. The mercurial vapours are conducted into six condensing chambers covered with iron plates and kept cool by a stream of water. The draught through the chambers is maintained by a chimney, where terraces are constructed over which water is allowed to flow. Fresh charges of 7 cwts. of ore and 28 lbs. of charcoal are introduced every three-quarters of an hour, and the spent ore is raked out from below through the moveable bars of the grate.

A still more scientifically constructed apparatus is employed at Idria for the treatment of small ores containing one-hundredth or less of mercury, in which the ore is heated on the hearth of a reverberatory furnace (*a*, Fig. 99), the mercurial vapour being passed—1, through a condensing chamber (*d*); 2, through a wide sloping iron pipe (*f*) kept cool by the constant trickling of water over it, from the perforated gutter (*e*); 3, into a second condensing chamber (*g*); 4, through a second iron pipe (*h*), similar to the first, into a chimney (*k*). The hearth of the reverberatory furnace is divided into three compartments, the ore being first

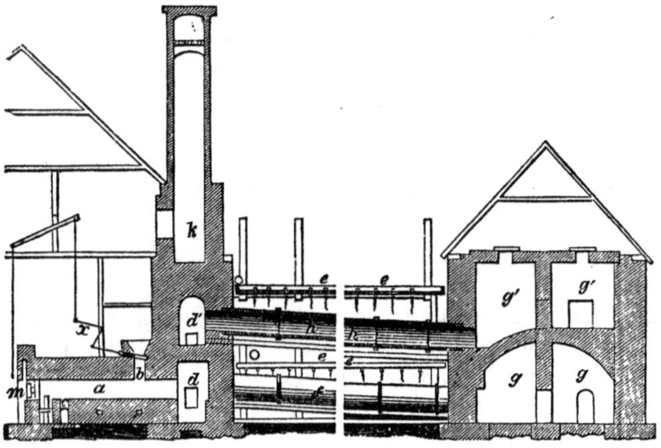

FIG. 99.—Modern Idrian Mercury-furnace.

thrown, through a hopper (*b*), upon that farthest from the grate, being raked out into each of the other compartments in succession, so as to be exposed to a gradually increasing temperature until it is drawn out in a spent condition through a channel near the fire-bridge, whilst fresh ore is charged at the other end of the hearth. Wood is the fuel employed. Beside the liquid mercury which is run out into proper receptacles, a large proportion of mercurial soot, containing finely-divided mercury, is collected in the condensing pipes and chambers. This is dried, raked over on an inclined plane as long as any mercury runs out, and afterwards distilled to obtain the last portions of the metal.

On the west bank of the Rhine there are several small mercury mines which yield sulphuret of mercury associated with sandstone. These ores are made to yield their mercury by distilling them with lime, a process generally resorted to in the smaller mercury works. The distillation is effected in cast-iron pear-shaped vessels (A, Fig. 100), thirty of which are heated by the same fire, in a *gallery furnace* (M), the grate of which runs through its entire length, and is fed with coal, which does not come into contact with the iron vessels, these being arranged in the upper part of the furnace, on each side of the grate, in two rows, one above the other, so that the flame may circulate around them before escaping through the openings into the chimney. The ground ore is mixed with about one-fourth of its weight of quicklime, and about 70 lbs. of the mixture are introduced into each of the cast-iron bottles, which are then about two-thirds full. The neck of each bottle fits into a stone bottle (B) half full of water, placed outside the furnace, to receive the mercury. The water above the metal becomes filled with *black mercury* containing undecomposed sulphuret and finely-divided mercury; this is dried and distilled again with more lime.

FIG. 100.—Extraction of Mercury in the Palatinate.

In this process, the lime, or oxide of calcium, composed of calcium and oxygen, decomposes with the sulphuret of mercury, yielding sulphuret of calcium, which remains in the iron bottle, mercury which passes over in vapour, and oxygen which converts a part of the sulphuret of calcium into sulphate of lime.

This operation in the gallery furnace is obviously attended with waste and inconvenience. The trouble of charging and discharging so many small bottles and of cementing the joints is very considerable, and has led to the introduction, in some places, of another arrangement which allows the mercury to be extracted on the same principle but with much greater economy.

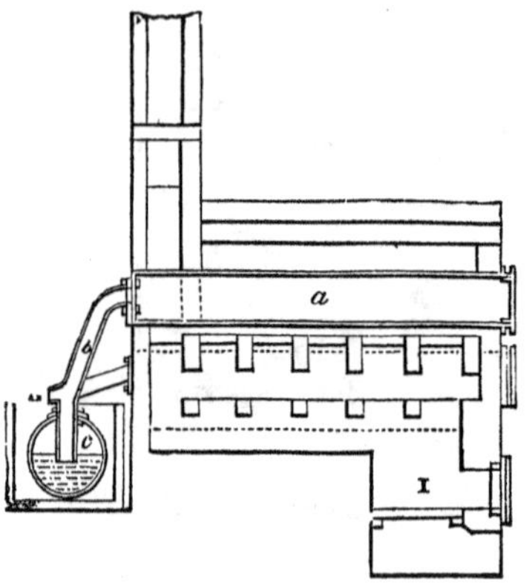

FIG. 101.—Ure's Retort for extraction of Mercury from Cinnabar.

In place of the bottles, cast-iron retorts (*a*, Fig. 101) are employed, resembling those used in distilling coal for gas. These are about seven feet long and one foot square in sectional area, so that they may be charged with 700lbs. (instead of 70) of the mixture of the ground cinnabar with lime, introduced at the back of the retort, which is then closed with an iron plate. The front end is also closed with an iron plate (*a*, Fig. 102), which is provided with a sloping cast-iron pipe (*b*), 4 inches in diameter, having a door through which it may be cleared with a wire. This pipe dips into water contained in a condenser (*c*) resembling the *hydraulic main* of the gas works, being an iron pipe, 18 inches wide, and 20 feet long, which runs along the front of the range of nine retorts, and receives the mercury condensed from them, being itself kept cool by a stream of water running through a wooden trough around it. This pipe is a little inclined towards one end, so that the condensed mercury may run down into a pipe (D) which conveys it into a locked cistern (*e*) (to prevent pilfering), in which an iron float (*k*) indicates the level of the mercury. The end of the pipe in the cistern is made to open into a small vessel, which is filled with mercury at the commencement to prevent the water from running out of the condenser. The latter is provided with a safety valve (*g*) to allow for any sudden expansion or contraction of the air within. The retorts are set like gas retorts, so that the same coal fire (I) may heat three of them, the range of nine having three furnaces with flues running into one chimney.

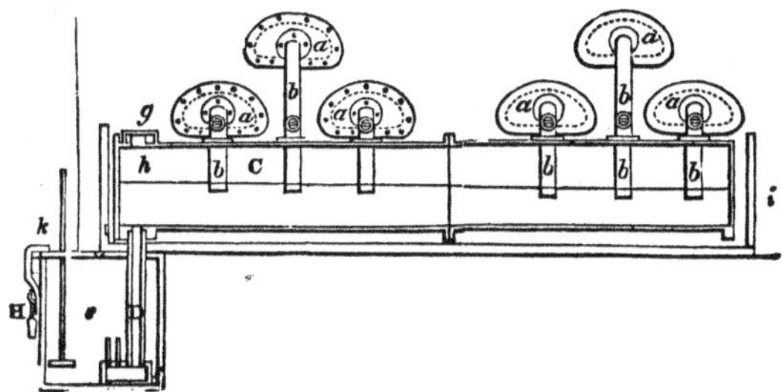

FIG. 102.—Ure's Retorts for extraction of Mercury.

In Hungary, a considerable quantity of mercury is extracted from the grey copper ore (*Fahl-erz* or *fallow ore*) during the process of roasting preliminary to the smelting for copper. The roasting is effected by burning the ore in mounds about 40 feet by 20, and 4½ feet high, in which channels are constructed for the admission of a moderate supply of air. A layer of small ore having been spread upon the ground, and covered with some larger pieces of ore already once roasted, some wood and coal are spread over them, then a layer of ore, the outside being covered with powdered ore so as to prevent a too free passage of air, and to retain the vapour of mercury. The heap is kindled through shafts left for the purpose, when the heat evolved by the combustion of the wood and coal ignites the sulphur of the ore, which burns away as sulphurous acid, whilst the mercury is converted into vapour and condensed among the cooler portions of the fine ore on the outside of the heap. After about three weeks, the upper layers are thrown upon a sieve and washed with water, the fine ore which passes through the sieve being separated from the mercury by washing.

The mercury imported into this country generally contains lead, bismuth, and zinc as impurities, which cause globules of it to *tail* or leave a metallic streak behind them when rolled over a glass plate, whilst pure mercury runs off in globules, leaving the glass clean.

It is sometimes purified by re-distilling it, using as a retort one of the iron bottles in which it is imported; but the metal is liable to violent concussions during the ebullition, so that it is impossible to prevent a part of the impure mercury from splashing over into the receiver. A far better process for purifying it consists in pouring the mercury into a wide dish (such as a photographic tray), where it may form a thin layer, and covering its surface with diluted nitric acid (sp. gr. 1·15). The acid is allowed to remain in contact with it for a day or two, being frequently stirred, until the mercury no longer tails upon glass. The lead, bismuth, and zinc are dissolved by the acid, in the form of nitrates, together with a portion of the mercury, so that the acid should be preserved in order to assist in the purification of another portion of the metal. The mercury may be separated from the acid by pouring them both into a funnel stopped with the finger, on removing which the mercury runs off, the finger being replaced to retain the acid; the mercury is then well washed with water in a dish, and dried, first with blotting-paper, and afterwards at a gentle heat. Any dust or other mechanical impurities may be easily removed by filtering the mercury through a cone of writing-paper, in the apex of which a few pin-holes have been made.

An old and simple process for the purification of a small

quantity of mercury, consists in shaking it in a bottle with air and a little powdered sugar, which helps to divide the mercury and expose the lead, &c., to the action of the air, which converts them into a grey powder. By repeating the agitation with fresh portions of air, nearly all the foreign metals may be removed, and the sugar with the adhering oxides may be separated by filtering through pricked paper.

Many of the useful applications of mercury depend upon its being the heaviest substance which is liquid at the ordinary temperature. Its specific gravity being 13.54, a column of mercury thirty inches high serves, in the barometer, to measure the pressure of the atmosphere, whilst thirty-three *feet* (396 inches) of water are required for the same purpose.

The great interval between the temperature at which mercury congeals to the solid state (71 degrees below the freezing point of water) and that at which it boils (450 degrees above the boiling-point of water) renders it especially suitable for filling thermometers, an application for which it is also recommended by the circumstance that it has a very low *specific heat*, or requires much less heat to raise it to a given temperature than most other liquids which could be employed for thermometric purposes: thus, a spirit thermometer, in which alcohol is used to show the expansion, rises and falls much more slowly than the mercurial thermometer, because the specific heat of alcohol is 15 times as great as that of mercury; in other words, fifteen seconds would be required by a spirit-thermometer to measure a temperature which would be indicated in one second by a mercurial thermometer of the same weight. A further advantage on the side of mercury is derived from its not adhering to the glass. Moreover, in a spirit-thermometer, the indications are affected by the presence of the vapour of alcohol in the space above the column, which should be perfectly vacuous, and is nearly so in the case of the mercurial thermometer, for the metal does not evolve an appreciable amount of vapour at temperatures below the boiling-point of water.

That mercury does give off a minute quantity of vapour, even at the ordinary temperature, is shown by the appearance of minute globules of metal condensed, in cold weather, upon the glass in the upper part of a barometer, and by the experiment of suspending a gold leaf at an inch or two above the surface of mercury in a bottle, when it is found, after some time, to be whitened by the combination of mercury vapour with the gold.

Mercury is remarkable for its property of uniting with other metals at the ordinary temperature, to form combinations which are termed amalgams. Iron and platinum are the only metals in ordinary use which are not attacked by mercury. It adheres to platinum and wets its surface, but does not unite chemically with it. Gold is soon penetrated by mercury, and becomes very brittle. The surface of a gold ring, for instance, is instantly whitened by mercury, and if allowed to remain in contact with it for a short time, the gold is rendered so brittle as to be useless; a mere external coating of amalgam may be removed, and the colour of the gold restored, by warming it with a little nitric acid, the surface of the gold being afterwards burnished.

Silvering Looking-glasses.—An amalgam of tin is employed for silvering the backs of looking-glasses, which is performed in the following manner. A sheet of tin-foil (generally hardened by a minute proportion of copper), somewhat larger than the plate to be silvered, is laid upon a stone or marble table, which is swung upon an axis so that it may be gradually sloped by a screw when required. The upper surface of this table is perfectly smooth and level, and has a gutter running round it, with a spout for collecting the superfluous mercury. The tin-foil is applied to the table with a brush, so that it may be free from wrinkles, and the surface having been laid truly horizontal, a little mercury is spread over it with a roll of flannel, so that every part of the tin-foil may be amalgamated. A very thin layer of mercury is then poured over the surface, and the edge of the clean dry plate to be silvered is very carefully pushed forward over the table, so as to carry the superfluous mercury before it, and to prevent any air from entering between the amalgam and the glass. Some flannel is placed on the glass, with a weight upon it, and the table is very slightly inclined to drain off the excess of mercury. After about five minutes, the plate is loaded with several heavy weights, and allowed to remain for twenty-four hours, in order that the amalgam may be made to adhere firmly to the glass; the inclination of the table is somewhat increased from time to time, to promote the draining away of the excess of mercury. When the weights are removed, the plate is laid upon a sloping wooden table, the upper edge of which is raised gradually by a pulley until the plate is perpendicular. After three or four weeks, when the excess of mercury has drained off, the looking-glass is ready for framing. The amalgam adhering to the glass contains one part of mercury and four parts of tin.

The amalgam which is employed to promote the action of glass electrical machines is composed of two parts of zinc and five parts of mercury.

Magnetic amalgam is the somewhat fanciful name bestowed upon the amalgam of one part of metallic sodium with thirty parts of mercury, which is liquid at a very moderate heat, but solidifies on cooling to a hard crystalline mass. It is cast into ingots which are kept in air-tight iron vessels with lime, to absorb any moisture, which would act upon the sodium in the amalgam and convert it into soda. This amalgam is exported to the amalgamating mills for gold and silver ores, where it encourages the amalgamation and prevents *flouring* (p. 221).

An amalgam of one part of cadmium and three parts of mercury is employed by dentists.

PLATINUM.

This valuable metal has been brought into use in quite modern times, having been discovered by an assayer in Jamaica in 1741, becoming generally known in Europe in 1748. Its name, derived from the Spanish, signifies *little silver*, since it somewhat resembles that metal in colour; it has also been called *white gold*, and it is said that when first discovered, much of it was thrown away, lest, from its durability and high specific gravity, it should be employed for debasing gold, an expectation which has been partly realised (p. 253).

Platinum always occurs, like gold, in the metallic state, and most commonly in alluvial deposits in which gold is also present. Nuggets of platinum have rarely been found, the largest on record being one of 18 lbs. weight. The metal is commonly met with in flattened grains of a light steel-grey colour. The Ural mountains have furnished the largest quantity of platinum, but it has also been obtained from Brazil, Peru, Borneo, Australia and California.

The *Platinum ore*, as it is called, is separated from the earthy matters by washing the latter away, when the grains of platinum remain behind with grains of gold, magnetic iron ore, corundum, and a very heavy alloy of osmium and iridium. The platinum itself is far from pure, containing only from 75 to 85 parts of that metal in a hundred, the remainder consisting of iron, sometimes found to the amount of 13 per cent., iridium, rhodium, palladium, osmium, and copper. When any considerable quantity of gold dust is present, it is separated from the platinum by the process of amalgamation. The magnetic iron ore is extracted by a magnet.

The original process for extracting platinum from the ore is rather a chemical than a metallurgical operation, but since it was, until within the last few years, the only method by which the metal was produced in a marketable state, a short description of it is here given.

The grains of platinum ore are heated with nitric acid, which dissolves any silver, copper, iron, and lead, in the form of nitrates; after these have been extracted, the residue is washed with water, and heated with hydrochloric acid, which dissolves the magnetic oxide of iron; it is then again washed with water, and gently heated for several hours with hydrochloric acid to which a little nitric acid is added from time to time. The platinum is thus dissolved, together with palladium, rhodium, and some iridium, whilst the osmium and the rest of the iridium are left undissolved. The acid solution containing the chlorides of platinum, &c., is poured off, and the residue heated with fresh portions of acid as long as anything is dissolved, when any quartz and corundum are left, together with the grains of the alloy of osmium and iridium, which are employed, on account of their surpassing all other metallic substances in hardness, for making the nibs of gold pens upon the points of which they are soldered.

The liquid containing the chloride of platinum is mixed with a solution of sal-ammoniac (muriate of ammonia) containing one-sixth of its weight of the salt, of which about four parts are employed for every ten parts of the ore. A yellow precipitate is then deposited, which contains the greater part of the platinum, in the form of *ammonio-chloride of platinum*, a combination of sal-ammoniac with chloride of platinum.

This yellow precipitate is washed with cold water, dried, and strongly heated in a plumbago crucible, when the sal-ammoniac and the chlorine are driven off, and *spongy platinum* is left as a grey porous mass. This is finely powdered in a wooden mortar, rubbed to a paste with water, passed through a sieve in order to render it perfectly uniform, and poured into a slightly conical brass mould closed below with blotting-paper wrapped round a steel stopper. When the water has drained off, a plunger is forced in by a coining press, so as to condense the mass, which has at first the specific gravity 4·3, until its specific gravity is 10, when it has been reduced to about two-fifths of its former bulk, and has acquired a metallic appearance. The disk is now sufficiently coherent to be removed from the mould and intensely heated for about 36 hours in a porcelain kiln, when it contracts to about four-fifths of its former volume. It is taken out of the furnace at a white heat, and hammered upon its ends, not upon the sides, lest it should crack. After being heated and hammered in this way several times, the particles become thoroughly welded together into a compact malleable mass of metal, of specific gravity 21·5.

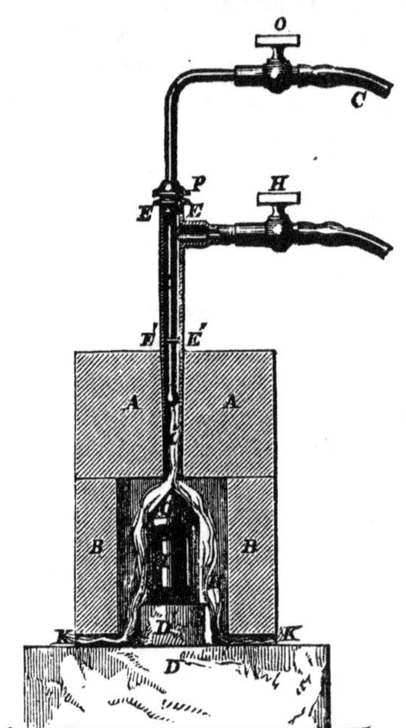

Fig. 103.—Platinum melted in Crucible by the Oxy-hydrogen Blowpipe.

Instead of welding the platinum into a compact mass, it is now sometimes melted in a crucible by the intense heat of the oxy-hydrogen blowpipe-flame. The crucible is made of gas-carbon, and is enclosed in another crucible (I, Fig. 103) made of lime, and provided with a conical cover (G) of the same material. The furnace is made of three blocks of lime strengthened by iron wire, one (D) serving for the hearth, upon which is placed a hollow cylinder (B) of lime, which has been bored in the lathe until it is wide enough to leave a clear space (H) of about $\frac{1}{6}$th inch round the lime crucible. At the lower part of this cylinder are four openings (K) for the escape of the steam produced in the combustion of the hydrogen and oxygen. The furnace is covered in with a block of lime (A) about 2$\frac{1}{2}$ inches thick, in which a slightly conical passage is bored for the reception of the blowpipe-jet which passes down to within an inch of the apex of the conical cover of the lime crucible. The blowpipe (E) consists of two concentric copper tubes, the outer one through which the hydrogen is passed being about $\frac{1}{2}$ inch in diameter, and terminating in a somewhat tapering nozzle of platinum (C) about 1$\frac{1}{2}$ inch long, which fits into the conical passage through the upper block of lime. The oxygen is conveyed through the inner copper tube, also furnished with a platinum nozzle, the opening of which is about $\frac{1}{10}$th inch in diameter.

The hydrogen and oxygen are supplied from their respective gas-holders, their passage being regulated by the stopcocks H O, under a pressure of about 16 inches of water, the hydrogen being lighted before the oxygen is turned on.

The new process for extracting platinum from its ores resembles one of the methods of extracting gold and silver, the metal being dissolved out by melted lead and afterwards recovered by cupellation.

Two hundredweight of the platinum ore mixed with an equal weight of galena (sulphuret of lead) is thrown, in small portions, into the concave hearth of a small reverberatory furnace built of fire-brick. The materials are stirred with an iron rod until the platinum has entirely dissolved in the fused galena. A little glass is then introduced, to melt over the surface, and a quantity of litharge (oxide of lead) equal in weight to the galena is gradually added. The sulphur in the sulphuret combines with the oxygen of the oxide, and passes off as sulphurous acid gas, leaving the metallic lead in combination with the platinum, whilst the alloy of osmium and iridium, being unaffected by the lead, sinks, in separate grains, to the bottom (its specific gravity varying, according to its composition, from 19·4 to 21·1). After remaining at rest for some time, the upper portions are ladled out and cast into ingots, and the remainder is added to the next charge.

The lead containing platinum is treated in a cupellation furnace (p. 208), when the lead is removed as an oxide,

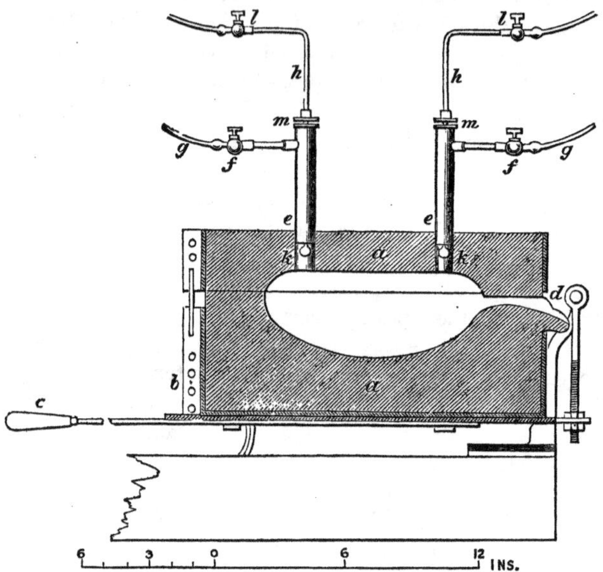

FIG. 104.—Lime Furnace for melting Platinum with the Oxy-hydrogen Blowpipe. *a a*, Blocks of lime hollowed out to form the furnace. *k k*, Openings for the oxy-hydrogen blowpipes. *e e*, Outer tubes conveying the hydrogen or coal-gas. *h h*, Inner tubes conveying the oxygen. *d*, Spout for pouring the melted platinum.

leaving the platinum in a spongy state upon the cupel, whence it is transferred to a small furnace made of lime (Fig. 104) and melted by the flame of the oxy-hydrogen blowpipe (in which coal-gas may be substituted for hydrogen), the intense heat of which volatilises any silver, gold, lead, palladium and osmium. When the metal is sufficiently refined, it is poured through an opening in the side of the furnace into an ingot mould made of gas-carbon or of wrought iron lined with platinum. In this manner, 25 lbs. of platinum have been melted and refined in three-quarters of an hour, with a consumption of about 43 cubic feet of oxygen. The melted platinum resembles silver in its property of absorbing oxygen mechanically at a high temperature and evolving it again as it cools, exhibiting the phenomenon of *sprouting* (p. 211).

Since platinum is one of the most malleable and ductile of the metals, being surpassed, in the former quality, only by gold, silver and copper, and in the latter by gold and silver, it is easily rolled into sheets or drawn into wire.

The principal uses of platinum depend upon its resistance to the action of heat, of oxygen, and of acids. The largest quantity of the metal is devoted to the manufacture of the stills employed for boiling down oil of vitriol in order to expel the water, and of the siphons used for drawing the hot acid out of the stills. Similar stills are also employed in the operation of parting gold and silver with sulphuric acid (p. 247). The joints of these stills are soldered with fine gold, and they are usually gilded inside, for otherwise they are liable to become porous under the influence of the boiling acid, allowing it to exude. They are protected from external injury and from the direct action of the fuel by an iron casing. Some platinum stills have been made weighing upwards of 60 lbs. and costing above 2,000*l.* They are slowly corroded by the action of the acid, and require occasional repair by soldering them with gold.

Platinum is also much employed for evaporating basins and crucibles for chemical purposes, and thin foil and wire of this metal are indispensable in operations with the blowpipe. When the particles of the metal have been imperfectly consolidated by hammering, it is found to blister under the influence of a very high temperature. The permanence of platinum under the action of heat has led some persons to adopt an erroneous estimate of its durability under other conditions, so that it is often forgotten that platinum is a soft metal and therefore ill-adapted to resist ordinary wear. It is easily corroded and rendered brittle by carbon and silica, both of which are present in coal, coke, and charcoal, for which reason platinum crucibles are never allowed to come into direct contact with the solid fuel, but are heated either in the flame of a gas or spirit lamp, or in a muffle (p. 255), or enclosed in a clay crucible lined with magnesia to prevent the platinum from sticking to the heated clay.

Metals must never be melted in platinum crucibles, since most of the metals are capable of forming alloys with it. Caustic alkalies and saltpetre in a melted state also act upon the metal, and phosphorus and arsenic combine with and corrode it very rapidly at moderately high temperatures.

Neither sulphuric, hydrochloric, nitric or hydrofluoric acid separately has any action upon platinum, but a mixture of hydrochloric with nitric acid dissolves it, though more slowly than it dissolves gold.

An alloy of platinum, iridium and rhodium is sometimes employed for crucibles, which are harder and less easily corroded than those made of pure platinum. To obtain the alloy, the ore of platinum (which contains the two other metals) is mixed with a quantity of lime equal to that of the iron contained in the ore, and fused by the oxy-hydrogen blowpipe in a furnace made of lime (p. 280). The iron and copper are converted into oxides which form a fusible slag with the lime, whilst the gold, palladium and osmium are expelled in the form of vapour, and the alloy of platinum, iridium and rhodium remains.

Small tubes, &c., may be easily extemporised with platinum wire and foil, by taking advantage of the readiness with which surfaces of this metal unite when hammered at a high temperature.

Platinum vessels are cleaned by smearing them with a paste containing equal bulks of borax and cream of tartar with a little water, drying and heating them till the mixture melts, and immersing them for several hours in diluted sulphuric acid. Heating in contact with fused bisulphate of potash, or with powdered sal-ammoniac, is also employed for the same purpose. The platinum vessels are finally well washed with water and burnished with agate.

Platinum is sometimes employed for the touch-holes of small-arms, and for the vents of cannon, on account of its resistance to corrosion. The circumstance that it expands less than any other metal when heated, enables it to be cemented into glass, by fusing the latter, whilst other metals which differ much from glass in their rate of expansion by heat, would crack it as they cool. This renders platinum of great importance in the fabrication of various philosophical instruments.

Though pure platinum is unaffected by nitric acid, it may be rendered soluble in that acid by previously alloying it with ten or twelve times its weight of silver, which is taken advantage of in order to separate platinum from gold in the process of assaying the latter metal with which platinum is frequently associated. If the platinum be present in small proportion (not exceeding 3 or 4 per cent.) in the alloy of gold and silver obtained by cupellation (p. 255), the whole of it will be dissolved together with the silver, in parting by nitric acid; but when the quantity of platinum is larger, which is indicated by the difficult fusibility of the button on the cupel, and by the blanched appearance of the gold eventually obtained, the latter must be again fused with at least three times its weight of pure silver, the alloy rolled very thin, and boiled for half an hour, and a quarter of an hour, respectively, with the two strengths of nitric acid mentioned at page 255, in order to remove the whole of the silver and platinum. An alloy of silver with one-third of its weight of platinum is employed by dentists on account of its great elasticity.

The remarkable property of platinum, especially in the

finely-divided states of spongy platinum and platinum black, to condense gases into its pores and thus to promote their chemical action upon each other, is not suited for description in a metallurgic treatise.

PALLADIUM is generally found in small quantity, not exceeding 1 per cent., associated with the ore of platinum, from which it is extracted by a process which is purely chemical. Formerly there existed a pretty abundant source of this metal in the form of an alloy with gold found in the mines of Brazil, but of late years this has failed, and palladium has risen to an extremely high price. In appearance it resembles platinum, but is much harder, though it possesses considerable malleability and ductility. It is quite unchanged by air at the ordinary temperature, but assumes a bluish colour, from the formation of a thin film of oxide, at a moderately high temperature, becoming bright again at a higher temperature, the oxide being decomposed. Palladium fuses at a somewhat lower temperature than platinum, but cannot be fused in a furnace. It is only half as heavy as platinum, its specific gravity being 11·5, so that it is much better adapted for making very accurate balances and other philosophical apparatus. The graduated scales of astronomical instruments are often made of palladium, and an alloy of this metal with $\frac{1}{10}$th of silver has been sometimes employed by dentists.

ANTIMONY.

Though antimony is far too brittle to be employed in its pure state for any useful purpose, it has been shown to be of great service in hardening the softer metals lead and tin, so that the history of this metal is not devoid of interest for the metallurgist.

Antimony is occasionally found in Nature in the metallic state, as at Andreasberg in the Hartz, where it is alloyed with small quantities of silver, iron, and arsenic. The only ore from which it is largely extracted is the *grey antimony ore*, a sulphuret of antimony, containing, when pure, $71\frac{3}{4}$ parts of antimony combined with $28\frac{1}{4}$ parts of sulphur. It is found in Cornwall, Auvergne, Hungary and Borneo, associated with galena and iron pyrites, and with quartz and heavy spar, in veins traversing rocks of granite or slate. The appearance of grey antimony ore is very characteristic; it commonly resembles a compact bundle of dark grey metallic needles converging towards one point, and often exhibiting a blue iridescence due to a thin film of oxide. It is very heavy (sp. gr. 4·63), and melts easily even in the flame of a candle. This fusibility is taken advantage of in order to separate it from the earthy matters, to effect which the ore is heated on the concave hearth of a reverberatory furnace, the hearth being lined with charcoal to prevent oxidation of the sulphuret, which melts and is run out into moulds, where it is cast into the form of the cakes sent into commerce as *crude antimony*, which contains, in addition to the sulphuret of antimony, sulphurets of arsenic, iron and lead.

The *oxide of antimony* occurs in Algeria, and is smelted in France.

Red antimony, a compound of oxide and sulphuret of antimony, is found in Tuscany, and smelted at Marseilles.

Regulus of antimony, or metallic antimony, is extracted from the sulphuret by melting it upon the hearth of a reverberatory furnace in contact with metallic iron (clippings from the tin-plate works), which removes the sulphur, forming sulphuret of iron; this collects above the melted antimony, which is run out into moulds. It contains a considerable quantity of iron.

Sometimes the regulus is extracted directly from the rich antimony ore, without previous production of crude antimony. For this purpose the ore is broken into pieces as large as an egg, and introduced into red-hot crucibles, together with a little alkaline slag; some scrap-iron is placed on the top and pressed down when the mass has fused. After about two hours, the melted antimony and sulphuret of iron are poured into conical iron moulds, where they separate into two layers.

A purer metal is obtained by the following process. The sulphuret of antimony, or the rich original ore, is crushed, and roasted, without being melted, for six hours, in a reverberatory furnace, when most of the sulphur is expelled as sulphurous acid gas, and most of the arsenic as arsenious acid, whilst part of the antimony is converted into vapour, and combines with oxygen to form the oxide of antimony, which is carried into the flues of the furnace.

The roasted ore, which has a red-brown colour, contains the oxide and sulphuret of antimony. It is ground to powder, mixed with about one-fifth of its weight of charcoal, some chloride of sodium, carbonate of soda, sulphate of soda, and slags from a former operation. The mixture is thrown upon the hearth of a reverberatory furnace, and well stirred, when the oxide of antimony in the roasted ore is reduced to the metallic state by the charcoal, whilst the sulphuret of antimony exchanges its sulphur for the oxygen of the soda, yielding oxide of antimony, which is also reduced by the charcoal, and sulphuret of sodium, which forms a slag with the sulphurets of other metals present, and with the chloride of sodium. The metal and slag are run off into an outer basin.

The fumes of oxide of antimony are condensed in long flues. The poorer ores, after being roasted, are smelted in cupola furnaces with coke.

The antimony is refined by melting it, in quantities of 60 or 70 lbs., with 1 or 2 lbs. of American potashes (carbonate of potash) and 10 lbs. of the slag. It is then allowed to solidify quietly under a layer of slag, in order that it may assume the beautiful fern-like crystalline markings on its surface which have gained for it the name of *star-antimony.*

The above process for extracting the metal is far from economical, little more than one-half of the antimony present being obtained in the metallic state.

Antimony can easily be distinguished from every other metal by its hardness, brittleness and crystalline structure; a slight tap with a hammer suffices to break an ingot of antimony, and the broken surface exhibits large shining plates;* it is so brittle that it may be easily reduced to a fine powder in a mortar. It is comparatively a light metal, its specific gravity being only 6·7. It melts at 800° F., and at a higher temperature it gives off much vapour, which produces a thick white smoke of oxide of antimony. Its applications have been noticed in the preceding pages.

* The addition of a minute proportion of tin to antimony causes it to crystallise more readily and in larger crystals.

BISMUTH.

Bismuth, or *marcasite** as it was formerly termed, is a comparatively rare metal which is found associated with the ores of nickel, cobalt, copper and silver, chiefly in Saxony, Transylvania and Bohemia. It also occurs in smaller quantity in Cornwall, Cumberland, Stirlingshire, Norway, Sweden, and the United States, and has lately been found in Peru. Bismuth is always extracted from the ores which contain it

FIG. 105.—Extraction of Bismuth.

in the uncombined metallic state, by taking advantage of the readiness with which it fuses (507° F.) and drains away from the other constituents of the ore. It is extracted chiefly at Schneeberg in Saxony, from an ore containing from seven to twelve parts of bismuth in a hundred, associated with a compound of arsenic and cobalt. This is broken into pieces about the size of a nut, and introduced, in charges of about 50 lbs., into sloping cast-iron cylinders (Fig. 105) heated by a wood fire. The lower opening (*b*) of each cylinder is closed with a fire-clay stopper having an aperture through which the melted bismuth may run out. The upper open-

* A term sometimes applied also to iron pyrites.

ings are closed by an iron door (*d*); in about ten minutes after charging, the metal begins to run into an iron pot (*a*), kept hot by a separate fire in the channel (K), and containing a little coal-dust to prevent the oxidation of the bismuth. The ore is raked about occasionally, to promote the separation of the metal, which is exhausted in the course of half an hour, when the residue is raked out from the upper end of the cylinder, where it falls, down an incline, into a trough of water, and a fresh charge is introduced. With five cylinders over a single furnace, one ton of ore is smelted in eight hours. The iron pans are emptied into moulds in which the bismuth is cast into bars weighing from 25 to 50 lbs. each.

At Joachimsthal, the bismuth ores have of late been smelted in crucibles, by melting them with about a fourth of their weight of scrap-iron (to combine with the sulphur), half their weight of carbonate of soda, to convert the silica into a slag of silicate of soda, one-twentieth of lime and the same weight of fluor spar. One hundredweight of the mixture is introduced into each crucible, and a part of the carbonate of soda is employed to cover the mixture. The crucible is closed with a lid, and strongly heated in a furnace. When the mass is pasty it is well stirred in order to effect perfect mixture, and, after complete fusion, it is ladled out into conical iron moulds, when the bismuth subsides into the narrower part of the mould.

As it comes into the market, bismuth contains considerable quantities of arsenic, iron and silver, which do not, however, seriously interfere with its limited applications. A great part of the arsenic is sometimes expelled by heating the metal in crucibles, filled up with charcoal to prevent access of oxygen, when the arsenic passes off in vapour.

When a sufficient quantity of silver is present, it is extracted from the bismuth by the process of cupellation (p. 208), the bismuth becoming converted into an oxide, which is removed, like the litharge formed when lead is cupelled, leaving the silver upon the cupel. It is stated that the great rise in the price of bismuth during the last few years is partly due to the circumstance that large quantities of it have been bought up and cupelled for the sake of the silver which it contains. When lead containing bismuth is cupelled, the lead is first converted into oxide, so that towards the end of the process, nearly pure bismuth is left on the cupel, which will oxidise in its turn if the process be continued.

Bismuth is known by its peculiar reddish colour and its highly crystalline appearance. A blow with the hammer breaks it easily, though it is not quite so brittle as antimony. Its tendency to crystallise is very remarkable; by melting some bismuth in a ladle, allowing a solid crust to form upon the surface, piercing this with two holes, and pouring the liquid bismuth out of one of them (the other allowing the air to enter), a mass of beautiful cubical crystals of bismuth is obtained.

Its brittleness renders bismuth unfit for use in the metallic state, by itself, except for the construction of *thermo-electric piles*, which are made of alternate bars of bismuth and antimony, and are employed as very delicate thermometers, in order to measure slight differences of temperature by the electric currents which they produce.

The uses of bismuth in alloys depend upon its low melting-point (507° F.), and its property of expanding very considerably during solidification, the solid metal occupying $\frac{1}{32}$nd more space than the liquid metal, affording an exception to the general law, which requires that cooling should produce contraction. Another remarkable physical peculiarity of bismuth is the circumstance that the specific gravity of the metal is diminished instead of being increased by strong pressure. Thus, a cylinder of bismuth of specific gravity 9·783 was placed in a steel cylinder fitted with a plunger, and subjected to a pressure of 200,000 lbs., when its specific gravity was found to have diminished to 9·556, having been reduced by about $\frac{1}{43}$rd part. The compressed bismuth exhibited scarcely any crystalline structure.

Newton's fusible alloy is composed of two parts of bismuth, one of lead, and one of tin, and melts at 201° F., so that it liquefies readily in boiling water, although the most easily fusible of its constituents, the tin, has a melting-point of 442° F. Such an alloy is used as a soft solder by pewterers.

Some kinds of type-metal and stereotype-metal contain bismuth in order that they may expand into the finest lines of the mould during solidification. For a similar reason, an alloy of tin, lead, and bismuth is employed for testing the finish of a die.

Bismuth is more easily converted into vapour than many other metals, and may be boiled at a moderate white heat.

ALUMINUM.

This metal, which is now often called *Aluminium*, although discovered by Wöhler in 1828, has only within the last few years been found capable of useful application in its metallic form. Though never found as a metal in Nature, it is probably the most abundant of all metals in a state of combination, since it exists in every variety of clay (silicate of alumina), its quantity varying from twelve to twenty parts in a hundred. Another mineral containing aluminum is *kryolite*,* in which the metal is combined with sodium and fluorine, and forms 13 per cent. of the mineral, which is found in abundance in Greenland.

* So called from the Greek for *frost* on account of its resemblance to ice.

Aluminum is extracted from a particular variety of clay known as *bauxite*, which is found at Baux, near Arles, in the south of France; this mineral contains about one-third of its weight of aluminum, combined with oxygen (forming *alumina*), together with silica, oxide of iron, and water. At Newcastle, where the metal is extracted from bauxite, the following process is adopted:

The ground mineral is mixed with *soda-ash* (containing carbonate of soda and caustic soda) and heated in a reverberatory furnace, when the soda combines with the silica and alumina, forming compounds known as *silicate of soda* and *aluminate of soda*, whilst the carbonic acid is expelled in the form of gas. The mass, after cooling, is treated with water, which dissolves the aluminate of soda. This solution is mixed with enough hydrochloric (muriatic) acid to remove the soda, when the alumina is separated as a gelatinous precipitate composed of *hydrate of alumina*, a compound of alumina with water. This is mixed with common salt (chloride of sodium) and charcoal powder, to a stiff paste, which is made up into balls as large as an orange, very thoroughly dried, and strongly heated in earthen cylinders through which perfectly dry chlorine gas is passed.

The carbon of the charcoal combines with the oxygen of the alumina, escaping as carbonic oxide gas, whilst the aluminum unites with the chlorine to form the chloride of aluminum; the latter enters into combination with the chloride of sodium, producing a *double chloride of aluminum and sodium* which distils over and condenses to a solid salt. Ten parts of this salt are mixed with two parts of sodium in small pieces, and with five parts of kryolite or of fluor spar, to form a liquid slag which shall cover the surface of the metal. This mixture is thrown upon the red hot hearth of a reverberatory furnace, which is then immediately closed to exclude air. The sodium acts violently upon the chloride of aluminum, abstracting its chlorine and liberating the aluminum, which collects, in the melted state, beneath a layer of slag containing the chloride of sodium and kryolite. The metal thus obtained always contains silicon and iron in considerable quantity.

Aluminum is a white malleable metal about as hard as zinc, and fusing at a somewhat lower temperature than silver. It is remarkably light, having a specific gravity of only 2·5, and is unaffected by air; unlike silver, it is not even tarnished by air containing sulphuretted hydrogen. A bar of aluminum suspended from a string sounds like a bell when lightly struck. In manufacturing objects of ornament from aluminum, a solder is employed which contains ninety parts of zinc, six parts of aluminum and four parts of copper.

At present, the principal demand for aluminum in this country is for the manufacture of *aluminum-bronze* or *aluminum-gold*, which is an alloy of aluminum with nine times its weight of copper (see p. 175).

An alloy of silver with two-thirds of its weight of aluminum is used in France, under the name of *tiers-argent*, as a substitute for silver, being much harder than that metal and less than half the price.

Aluminum is sometimes employed for making small weights, for which it is well adapted by its lightness and resistance to the action of air. The beams of small balances have also been made of aluminum.

MAGNESIUM.

Like aluminum, this metal has only been extracted in any quantity during the last few years, a considerable demand for it having arisen in consequence of its property of burning with a very brilliant white light which is found useful for the illumination of microscopes, magic lanterns, &c., as well as for taking photographs at night or in places where daylight does not penetrate.

Magnesium occurs abundantly, in combination with oxygen and carbonic acid, in *magnesite* (carbonate of magnesia) and *dolomite* or magnesian limestone (carbonate of lime and magnesia). Another source of the metal is the recently-discovered mineral *carnallite*, which is found in large quantity above the rock-salt in the salt mines of Stassfurth in Saxony. This mineral is composed of magnesium, potassium, chlorine, and water, and contains about one-twelfth of its weight of magnesium. The water may be expelled by heat, leaving the *double chloride of magnesium and potassium.*

Magnesium may be extracted from the dried carnallite by mixing it with one-tenth of its weight of fluor spar, to act as a flux, and one-tenth of its weight of sodium in small pieces. By fusing this at a moderate heat, the chloride of magnesium is made to give up its chlorine to the sodium, and the magnesium collects in the melted state beneath a liquid slag composed of chloride of sodium, chloride of potassium, and fluoride of calcium. The magnesium may be purified by distilling it, in an iron crucible, as practised in the case of zinc (p. 156). Magnesium bears considerable resemblance to aluminum, but is a whiter metal, and even lighter than aluminum, its specific gravity being only 1·74. It may be liquefied below a red heat, and, as stated above, may be readily distilled. It is a little more tarnished than zinc when exposed to air. The magnesium wire is made by forcing the heated metal through holes in a steel plate, and magnesium riband, by passing the wire between heated rollers. When the end of a piece of wire or riband is held in a flame, it catches fire and burns with a dazzling light, the magnesium combining with the oxygen of the air to form a white earthy mass of magnesia.

The *sodium* required for the extraction of aluminum and magnesium is extracted directly from carbonate of soda, which is itself made from common salt (chloride of sodium). The well-dried carbonate of soda is mixed with powdered charcoal, some chalk being added to prevent the fusion of the mixture, which is strongly heated in wrought-iron cylinders protected from the fire by a coating of clay. The carbonate of soda contains sodium, oxygen, and carbonic acid; the carbon of the charcoal combines with the oxygen, and the sodium is converted into vapour and condensed in vessels containing petroleum; for sodium cannot be exposed to the air, even for a few minutes, without combining extensively with oxygen, and it even takes up that element, with great violence, from water, in which the oxygen is united with hydrogen. Sodium would scarcely be taken for a metal by an ordinary observer, in the state in which it is found in commerce, where it occurs in greyish earthy-looking light masses; but when these are cut with a knife, the fresh surfaces exhibit a brilliant lustre.

CADMIUM.*

Cadmium is found, in small quantities, not exceeding 2 or 3 per cent., in the ores of zinc, and distils over, together with the first portions of zinc, during the smelting of the ores of that metal, the period of its distillation being known as the *brown blaze*, because its vapour imparts a brown colour to the flame. If these first portions of metal, mixed with oxide, be mixed with charcoal and distilled again, the first portions, being collected apart, will contain a still larger proportion of cadmium, this metal being much more easily converted into vapour than zinc is. To obtain pure cadmium, the mixture of zinc and cadmium is dissolved in diluted sulphuric acid, and sulphuretted hydrogen gas is conducted into the solution, when the sulphur combines with the cadmium to form a bright yellow sulphuret of cadmium which is deposited. This is washed, dissolved in strong hydrochloric acid, and converted into carbonate of cadmium by adding carbonate of ammonia. The carbonate of cadmium being washed, dried, and distilled with charcoal, as in the case of zinc, yields metallic cadmium in a pure state.

Cadmium resembles tin in colour and appearance, as well as in the property of *creaking* when bent. It is a malleable and ductile metal at the ordinary temperature, but becomes brittle at about 180° F. It melts at the remarkably low temperature of 442° F., the melting-point of tin, and an alloy of three parts of cadmium, fifteen of bismuth, eight of lead, and four of tin fuses at 140° F.

Cadmium is harder than tin, and possesses greater tenacity. It is also somewhat heavier, its specific gravity being 8·6.

* From *cadmia* (Latin), *brass-ore*, referring to its connexion with the ores of zinc.

QUESTIONS FOR EXAMINATION.

General Properties of Metals (pp. 1–10).

1. What is meant by the *tenacity* of a metal? Compare tin, steel, and copper with respect to this quality.

2. Why are metals annealed during the process of rolling them into sheets?

3. Explain why iron, though less malleable than lead, is more ductile.

4. The specific gravity of mercury is given as $13\frac{3}{8}$; what does this signify?

5. Spirit of wine burning in an earthen saucer is extinguished when poured upon an iron plate; how is this accounted for?

6. If expense were no object, what metal would be best adapted for telegraph wires?

7. Which of the metals in common use can be melted in a ladle placed upon an ordinary fire?

Iron (pp. 11–103).

8. What is *meteoric iron*? By what metal is it always accompanied?

9. Which of the ores of iron are most abundant in England and Scotland, respectively?

10. Why are some iron ores calcined or roasted previously to smelting them?

11. By what considerations is the pressure of the blast from the twyers of a blast-furnace regulated?

12. Explain the principles upon which the flux for iron ores is selected.

13. Mention the chief chemical operations which take place in the blast-furnace.

14. What is the nature of the gas issuing from the blast-furnace, and how is it turned to account?

15. Point out the chief ingredients of the blast-furnace cinder. What does a black slag indicate?

16. What are the essential constituents of cast iron?

17. Explain the difference between *graphite* and *combined carbon*, and state how their proportions influence the properties of the cast iron.

18. What is the effect of casting in *chills*, and how can it be accounted for?

19. By what process can hard castings be softened externally?

20. Under what circumstances does the blast-furnace yield white iron?

21. Which of the varieties of cast iron contains the largest proportion of silicon?

22. What quality of cast iron is best suited for fine castings?

23. Why is the yield of iron from a given furnace greater in winter than in summer?

24. What is the mode of proceeding to obtain a mottled iron from a given furnace?

25. Explain the meaning of *hot-blast all mine* iron.

26. Describe a cupola furnace, and its use.

27. For what purpose is lime introduced into a cupola furnace? What evil results from using an excess of lime?

28. State how chilling may be avoided when casting in sand.

29. Describe the *finery-hearth*.

30. How is the composition of pig-iron affected by the process of refining?

31. Compare the composition of finery-cinder with that of blast-furnace cinder.

32. Explain the principle of the puddling process.

33. Mention the essential particulars in the construction of a puddling-furnace.

34. Point out the difference between *dry puddling* and *pig-boiling*.

35. State the composition of tap-cinder.

36. Mention the principal defects of the puddling process.

37. What are the characters of mill bar, and how is it converted into merchant and best bar?

38. In what particulars does the reheating-furnace differ from other furnaces?

39. What explanation can be given of the improvement of bar-iron by piling?

40. Explain the meaning of the terms *hot short* and *cold short*, as applied to iron.

41. By what physical properties is steel distinguished from iron?

42. What proportion of carbon is sufficient to constitute a mild steel?

43. Point out the best materials for making permanent and temporary magnets, respectively.

44. How may iron bars be converted into steel?

45. State the peculiar features of the cementation furnace.

46. What is the influence of the duration of the cementing process upon the character of the steel?

47. Point out the defects of blister steel, and describe its conversion into shear steel.

48. How is cast steel produced? What advantages are secured by adding carburet of manganese to it?

49. Mention some of the modern processes for casting steel ingots of large size.

50. Give a simple chemical test for distinguishing a steel blade from an iron blade.

51. Is any chemical difference observable between hard and soft steel?

52. By what process can the highest degree of hardness be conferred upon steel?

53. Why is oil sometimes employed for hardening steel?

54. Describe the process for reducing the hardness of steel.

55. What is the object in bluing steel articles, and how is it effected?

56. Describe and explain the process of case-hardening.

57. What descriptions of articles are usually case-hardened?

58. By what mode of treatment may small castings be converted into malleable iron?

59. What is meant by the *Catalan process* for the extraction of iron? Why is it not practised in England?

60. How can the specific gravity be employed as a test of the quality of iron and steel?

COPPER (pp. 103–130).

61. Where is copper chiefly found in the metallic state?

62. Give the composition of the most abundant of the English ores of copper.

63. Name the chief seat of British copper smelting, and state the reasons for its selection.

64. What is the composition of fluor spar, and why is it useful in copper-smelting?

65. Enumerate the chief stages of the process for smelting copper ores, with the objects to be attained by each.

66. Explain the management of the anthracite fire in the roasting-furnace.

67. What is the nature of *copper-smoke*?

68. Point out the chief differences between metal-slag and ore-furnace slag.

69. What advantage is secured by air-channels in the fire bridge of the roasting-furnace?

70. Give an explanation of the terms *underpoled* and *overpoled* copper.

71. What is the reason for scorifying copper with lead before rolling?

72. Why does the process adopted at Mansfeld differ from the Welsh process of copper smelting?

73. Describe the process for refining the rosette copper at Mansfeld.

74. How is cement copper obtained, and what is its quality?

75. What is bean-shot copper used for, and how is it prepared?

76. Is impure copper more inclined to hot-shortness or to cold-shortness?

77. Which of the uses of copper is most seriously interfered with by the presence of small quantities of foreign matters?

TIN (pp. 130–152).

78. What is the composition of tinstone, and where is it chiefly found?

79. Explain the meaning of *stream-tin*.

80. Give an outline of the mechanical treatment of tin ores.

81. Describe the smelting-process, pointing out the precautions against loss of tin in the slag.

82. How is tin refined by liquation?

83. What are the objects of *boiling* and *tossing*?

84. How does the process followed in the Saxon tin-works differ from the English process?

85. What useful compound is prepared from the wolfram associated with tin ores?

86. By what physical peculiarity may tin be immediately recognised?

87. How is tin-plate made?

88. Point out the difference between charcoal-plate and coke-plate.

89. Why is a kettle of block-tin so much more expensive than a common tin kettle?

90. How is copper cleaned previously to tinning it?

91. Should alloys be regarded as merely mechanical mixtures, or as chemical compounds?

92. Mention the chief useful alloys of copper and tin.

93. State the composition of the current bronze coin.

94. What is the method of hardening and tempering bronze?

ZINC (pp. 153–175).

95. Enumerate the chief ores of zinc, stating their composition.

96. In what respect does the process of extracting zinc differ from those employed for most metals?

97. What is the commercial name for zinc? Is any other substance ever called by it?

98. By what rather rare metal is zinc usually accompanied in its ore?

99. Under what conditions may zinc be rolled into sheets?

100. How can zinc be purified from lead?

101. What is meant by *burned zinc*?

102. *Corrugated iron* is used for building ; what is it?

103. What impurity is especially objectionable in the zinc employed for making brass?

104. State the composition of Dutch leaf gold.

105. For what reason is a little lead sometimes added to brass?

106. How is brass lacquered?

107. By what process are pins coated with tin?

108. State the composition of sterro-metal. For what purposes has it been employed?

109. What is aluminium-bronze?

LEAD (pp. 175–206).

110. Give the name and composition of the most abundant ore of lead.

111. What other metals are generally found in lead ore?

112. Point out the peculiar features of the reverberatory furnace for lead-smelting.

113. In what respect does the process of smelting galena resemble that adopted for copper-pyrites?

114. Why is iron sometimes added to the charge in the smelting-furnace?

115. In what respect does the economico-furnace differ from the reverberatory furnace for smelting lead ores?

116. What is the nature of the improving process?

117. How may a poor argentiferous lead be rendered fit for cupellation?

118. Point out the distinction between the high and low systems of concentrating silver-lead.

119. By what method can leaden pipes be protected from the corrosive action of water?

120. Of what material are metallic pencils composed?

121. State the composition of type-metal.

122. Why does antimony unfit lead for rifle-bullets?

123. How is the spherical form given to small shot?

124. Describe the mode of constructing the leaden chambers for the vitriol-works.

125. What is the nature of the change by which lead is converted into an earthy mass, easily crumbling?

126. State the composition of pewter. Why is the standard proportion of lead fixed at one-fifth only?

127. What is the difference between pewterer's solder and common solder?

128. Why is zinc dissolved in the muriatic acid which is used for soldering?

128A. Give the composition of hard solder. What is the effect of adding silver to it?

SILVER (pp. 206–238).

129. Explain the principles of the process of cupellation.

130. What is litharge? How is it obtained?

131. Why does melted silver sprout in solidifying?

132. Mention the principal differences between the English and German cupellation processes.

133. Describe the extraction of silver from the black copper at Mansfeld.

134. What is the principle of the extraction of silver by amalgamation?

135. Why is this process carried on so differently in Mexico and Saxony?

136. Describe a process in which common salt is employed for extracting silver from the ore.

137. What is the composition of silver coin? Why is not pure silver circulated?

138. What is understood by the assayer's statement that a sample of silver is *worse* 5 *dwts.*?

139. How can the genuine character of a shilling be tested without injuring it?

140. Why does silver tarnish so quickly where coal-gas is employed? How may it be cleaned?

141. Point out the difference between old silver plate and electro-plate.

142. What solution of silver is best adapted for electro-plating?

GOLD (pp. 239–261).

143. In what respect does the mode of occurrence of gold in nature differ from that of most other metals?

144. Describe the washing of auriferous sand in the cradle.

145. How is gold extracted from pyrites in the Tyrol?

146. Describe the process of parting by sulphuric acid. How is the silver recovered in the metallic form?

147. Is native gold absolutely pure?

148. What is meant by *fine gold*, and by *gold of* 18 *carats fine*?

149. By what rough test is gold commonly tried? Show where it is fallacious.

150. Explain the use of the touchstone.

151. What is meant by *inquartation* in the assay of gold, and why is it necessary?

152. Describe the manufacture of gold-leaf.

153. How are buttons coated with gold?

MERCURY (pp. 261–275).

154. In what forms is mercury met with in the mineral kingdom?

155. What is the nature of the process employed for extracting mercury at the works in Idria?

156. Is there any more economical process than this?

157. Give a ready method for ascertaining the purity of quicksilver.

158. Why is mercury selected for filling thermometers?

159. What is an *amalgam*?

160. How are looking-glasses usually made?

PLATINUM (pp. 276–284).

161. What is the condition of platinum in the ore?

162. Point out the chief differences between the old and new processes for treating platinum ores.

162A. How is platinum melted in considerable quantities?

163. In what branch of manufacture is platinum largely employed?

164. Why should platinum vessels never be heated in a coal fire?

ANTIMONY &c. (pp. 285–295).

165. What is the difference between crude antimony and regulus of antimony?

166. By what peculiarities is antimony distinguished from all other metals?

167. What is remarkable in the process of extracting bismuth from its ores?

168. Mention the impurities commonly found in the bismuth of commerce.

169. Upon what peculiarities do the uses of bismuth depend?

170. Where is aluminium met with? Does its high price depend upon its scarcity?

171. What useful applications have been found for aluminium?

172. From what circumstance does magnesium derive importance?

INDEX.

LONDON : PRINTED BY
SPOTTISWOODE AND CO., NEW-STREET SQUARE
AND PARLIAMENT STREET

www.ingramcontent.com/pod-product-compliance
Lightning Source LLC
LaVergne TN
LVHW020218110826
845151LV00003B/755

9781425533236